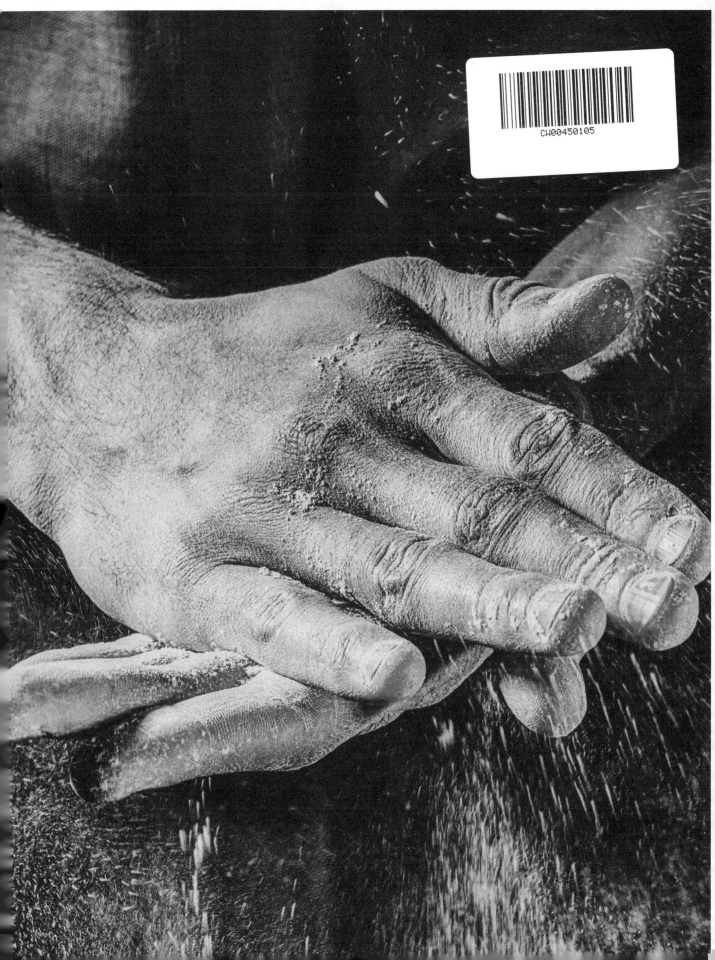

The Easy Bread Machine Cookbook

Easy Bread Machine Recipes to Save You Time. Simple recipes for beginners to Making delicious Homemade Bread

By

Michelle Gallagher

Table of Contents

Introduction

Bread has existed for more than 35,000 years, proof of which can be found in pottery and ancient caves worldwide. Archaeological evidence suggests that a wheat-based diet became a cornerstone of human existence nearly 10,000 years ago, and wheat remains the preferred grain to this day.

The first bread recipes appear around 1,500 B.C. in Egyptian cookbooks, and by the 7th century A.D. in Greece. The Roman invasion of Egypt, around 200 B.C., led to the spread of flatbread across Europe, where it was adopted in all its forms and became a staple food across the continent, especially in Italy.

In the fourth century A.D., when the Huns conquered Italy's northern part, the bread-baking profession became an exclusively male domain. When southern Italy was brought under Christian rule by the Normans in the late 11th century, the bread-baking secrets were brought to France, which today boasts some of the world's most incredible bakers. Baking was, and still is, an art not easily mastered, and the bread lessons handed down from generation to generation were passed on in the guarded ways, which only time-proven bakers could do.

When Napoleon conquered the Italian peninsula in the early 1800s, the bread-baking

industry was turned on its head and gradually moved south. Today, Italy is credited with being the largest bread-baking nation globally, but France remains the bread capital of Europe. However, advances in food production in the mid-1900s led to a ban on naturally leavened bread that is still in effect today throughout Europe, excluding a few small markets, where loaves must be labeled as such.

Throughout the 20th century, bread baking remained the domain of home-made artisanal bakeries that produced hundreds if not thousands of different types of bread, all unique and distinct from others, and known by their specific names: baguette, ciabatta, baton, and fougasse.

The domestication of wheat, and from there to bread, is a long, arduous, and intriguing path, as are practically all of man's discoveries. Bread has been an integral part of the diet of most cultures throughout history. It has been used as currency and barter, as food for religious rituals, and as a way of preserving grains in storage. Although today, it has mostly been replaced by the more convenient method of keeping through bottling, it still holds a place in most people's hearts.

But bread isn't that simple. It's not just about wheat and water. Mostly, it's about time. Bread has a particular soul. Bread depends on time to be right. In both, it is baking and eating; bread leaves a mark on the consumer that acknowledges that every grain of bread is a life cycle away from its final, highest point. It needs time to be right and to become something more.

What is the highest point of bread? Can there be higher points than the thousand delicacies we proudly display and frame on our bookshelves, the daily new taste we all know so well? The answer is no. Because now, new recipes are coming to replace our old-world ways. In bread, even the greatest of recipes do not define the final product. They serve almost a form of parallelizing, both bread and human evolution. New recipes open doors across the world. Inspired by time, they allow our bread, once plentiful and yet to be discovered, to be found through time.

Now, let's get down to business. It is the bread machine cookbook. It is something that will inspire you to look further aside from the ordinary. It is something that'll make your kitchen a place that friends, family, and passersby will wonder, "how'd they do that?" The recipes here will do you right. They will save you money and give you more time between the old you and the new you.

So, what exactly is a bread machine? It is a device that works as a regular bread machine but allows you to experiment with different ingredients. If you are a beginner, it will serve as a tool for learning the correct way of making delicious bread. You will find that the methods of making particular delicious bread are not that hard. What's great is that, unlike a regular bread machine, the ingredients for the recipes in this cookbook do not have to be bought online but have to be found in the nearest grocery store. And, as a bonus for a limited time, you can get your first recipe at absolutely no charge here. With all that said, try some new bread recipes with your bread machine, and happy bread making!

Chapter 1.

How to Bake Using a Bread Machine

Home bread makers are designed in such a way that any housewife can use them without much difficulty. However, for the stove to serve for a long time, and the bread always turns out to be high, lush, and tasty; specific rules must be followed.

You need to install the bread maker away from batteries, stoves, and sunlight since all temperature factors affect the oven's heating.

Before each new cooking, make sure that no crumbs are stuck on the blades, and that the edge is on the shaft until it stops.

When laying the components, you must strictly follow the instructions, for example: if you want to start with liquids; then first pour in water, milk, or other liquid product. Flour is passed to completely cover the wet layer, and then different dry ingredients are poured. Salt, sugar, hard butter (butter) are placed in the grooves made in the stacked layers so that they do not come into contact with each other. Then, in the middle of the layer of dry ingredients, the dough is pressed and poured (the depression should not reach the wet layer).

A food container is placed in the oven (there are usually special staples), the lid is closed, and the oven is plugged in. Select a program, the size of the final product, and the case (if provided on the model). Press the "Start" button. After that, the fermentation process begins. If your oven has a timer, you can set the bread making time for a specific time.

During kneading, the dough is checked by periodically opening the lid. To make good bread, the dough should be slightly sticky to the touch. If the dough is too soft and moist, add a little flour; if it turns out to be very dense, add liquid.

It is essential to assess the state of the dough during the lifting process. The dough may rise too high on hot days, and then it falls out of the mold and falls on the heating coils.

To not change the baking program, the dough can be punctured in several places to fall off. Or, cancel the originally specified program and set the mode, which in many models is called "Baking Only."

All additives fall asleep after the stove signal about the end of the kneading; also, by a timer indicating that the kneading process is completed. If the stove has an automatic addition mode, then all components are poured into a special compartment at the beginning of cooking, as we have already mentioned earlier.

At the end of the cycle, the bread maker beeps. It either turns off itself in automatic mode, or you should press the "Stop" button. After that, the lid is opened, and gloves are put on. The bread is then taken out (it is not recommended to lean close to the open stove and also rely on it).

Then turn the mold over the board, take out the bread, put it on the wire rack so that it cools down gradually. Then turn off the stove from the network and let it cool down (it is not recommended to start preparing a new portion of bread without waiting for the furnace to cool down).

It is recommended to use freshly baked bread for food within two to three days. It must be remembered that products containing eggs stale faster. Bread containing honey and butter retains its freshness and elasticity longer.

Chapter 2.

Why We Love Bread Machines

The main reason you will love your bread machine is the incredible variety of loaves you can create and enjoy without spending hours in the kitchen. Something is satisfying about popping out fragrant, fresh bread from the bucket, knowing you made it yourself. However, here are some other reasons your bread machine will quickly become your favorite kitchen appliance:

·Saves power.

Think about how warm your kitchen gets when you bake anything in the oven, especially during the summer months. Escaping heat is money out of your pocket, and even if you have a very well-insulated oven, it still costs more to run an oven than a bread machine. A standard bread machine's power usage is thought to be about the same as or less than a coffeemaker, about 9 kilowatt-hours for 15 hours a month.

·Set it and forget it.

You have time to do other tasks, run errands, or make the rest of a big meal without supervising the bread in the machine. Traditional bread-making is more hands-on and requires a significant amount of time.

·Cleanup is a breeze.

If you have ever made bread without a machine, you will be familiar with the dirty bowls, flour mess from kneading the dough, and washing up the pans or baking sheets. With the bread machine, you dump the ingredients in and wash a couple of measuring cups and spoons. Plus, bread machine buckets wipe off easily after the loaf is out.

·Control over ingredients.

Knowing what is in the food you set on the table is reassuring, especially if you have someone in the family with allergies or food issues. The ingredients you buy are completely under your control, so there are no unrecognizable ingredients in the finished loaf.

·Saves money.

All the ingredients you put into bread can be bought in bulk, which is the best way to save money.

The Glory of Gluten

Gluten is a protein composite found in cereal grains that produces the texture and structure of baked goods. Gluten is made up of two protein groups, gliadin, and glutenin, which bind together when mixed with water. The gliadin is responsible for gluten's ability to be stretched, and glutenin contributes to gluten's ability to return to its initial position after stretching. As the dough is mixed, a network of gluten strands is formed. The longer this process goes on, the stronger these strands become.

This gluten network traps the gases produced by yeast and allows the bread to rise. Some of the factors that affect gluten development include:

·Type of flour.

The protein content and quality of bread will vary depending on the type of flour.

·Amount of water.

Gluten doesn't exist without water.

·Fat.

Fat coats the proteins, blocking hydration and the formation of long gluten strands.

·Salt.

Salt makes gluten stronger and stickier.

·Sugar.

Sugar attaches to the water molecules before they can bind the protein groups together.

Bread machines are a fantastic kitchen accessory to own.

These small compact wonders have many options and settings for baking an assortment of bread masterfully. Once you become familiar with your bread machine's settings, the chance to create and experiment is endless.

Chapter 3.

Whole-Wheat Breads

Whole-Wheat Bread

Preparation:

9 minutes

Cooking Time:

4 Hours

Servings:

12 slices

INGREDIENTS

- 1 cup lukewarm water
- 2 teaspoon olive oil
- 1 2/3 cups whole-wheat flour sifted
- 1 tsp salt
- Soft brown sugar
- 1/2 cup dried milk powder, skimmed
- 0.25 oz. fast-acting, easy-blend dried yeast

DIRECTIONS

- Add the water and olive oil to your machine, followed by half of the flour.
- Now apply the salt, sugar, dried milk powder, and remaining flour.
- Make a little well or dip at the top of the flour. Then carefully place the yeast into it, making sure it doesn't come into contact with any liquid.
- Set the wholemeal or whole-wheat setting according to your machine's manual, and alter the crust setting to your particular liking.
- Once baked, carefully remove the bowl from the machine and remove the loaf, placing it on a wire rack to cool. I prefer not to add any toppings to this particular loaf, but you can, of course, experiment and add

whatever you want.

- Once cool, remove the paddle; and, for the very best results, slice with a serrated bread knife. Enjoy!

NUTRITIONS

- Calories: 160 Cal
- Carbs: 30.1 g
- Fat: 3,1 g
- Protein: 5 g

Whole-Wheat Peanut Butter and Jelly Bread

Preparation:

10 minutes

Cooking Time:

3 hours

Servings:

12 slices

INGREDIENTS

- 10 oz. of water at 90°F-100°F (320°C-370°C)
- ½ cup smooth peanut butter
- ½ cup strawberry jelly (or any preferable jelly)
- 3 Tablespoons vital wheat gluten
- ½ teaspoon salt
- ¾ teaspoon baking soda
- 1 ½ teaspoons active dry yeast
- ¾ teaspoon baking powder
- 1/4 cup light brown sugar
- 3 1/3 cups whole-wheat flour

DIRECTIONS

- As you prep the bread machine pan, add the ingredients following this particular order: water, jelly, salt, peanut butter, brown sugar, baking powder, baking soda, gluten, whole-wheat flour, and yeast.
- Choose 1 ½ Pound Loaf, Medium Crust, Wheat cycle, and then START the machine.
- Once baked, place it on a rack to cool and then serve.
- Enjoy!

NUTRITIONS

- Calories: 230 Cal
- Carbs: 39 g
- Fat: 6 g
- Protein: 9 g

Butter Up Bread

Preparation:

10 minutes

Cooking Time:

3 hours

Servings:

12 slices

INGREDIENTS

- 3 cups of bread flour
- ¼ cup margarine, melted
- 1 cup buttermilk at 110°F (450°C)
- 1 Tablespoon sugar
- 0.25 oz. active dry yeast
- 1 egg, at room temperature
- 1 teaspoon salt

DIRECTIONS

- Prepare the bread machine pan by adding buttermilk, melted margarine, salt, sugar, flour, and yeast in the order specified by your manufacturer.
- Select Basic/White Setting and press START.
- Once baked, transfer onto wire racks to cool before slicing.
- Enjoy!

NUTRITIONS

- Calories: 231 Cal
- Carbs: 36 g
- Fat: 6 g
- Protein: 8 g

Cracked Fit and Fat Bread

Preparation:

5 minutes

Cooking Time:

3 hours 25 Minutes

Servings:

16 slices

INGREDIENTS

- 1 ½ cups water
- 2 Tablespoons butter softened
- ¼ cup brown sugar
- 1 Tablespoon salt
- 3 cups bread flour
- 1 cup whole wheat flour
- 1 ½ cups cracked wheat
- 1 ½ oz. active dry yeast

DIRECTIONS

- In the bread machine pan, measure all components according to the manufacturer's suggested order.
- Choose Basic/White cycle, medium crust, and 2 lbs. weight of loaf, and then press START.
- Once baked, allow the bread to cool on a wire rack before slicing.
- Enjoy!

NUTRITIONS

- Calories: 65 Cal
- Carbs: 12.4 g
- Fat: 1 g
- Protein: 2 g

Easy Home Base Wheat Bread

Preparation:

10 minutes

Cooking Time:

3 hours 50 Minutes

Servings:

12 slices

INGREDIENTS

- 1 1/3 cup whole-wheat flour
- 1 1/3 cups bread flour
- 2 teaspoon butter softened
- 1 cup warm water at 90°F (32°C)
- 1 cup warm milk at 90°F (32°C)
- 0.25 oz. active dry yeast
- 1 egg, at room temperature
- 1 tsp. salt
- 1 2/3 Tablespoons honey

DIRECTIONS

- Add the ingredients into the pan of the bread machine following the order suggested by the manufacturer.
- Use the Whole Wheat cycle, choose the crust color, weight, and START the machine.
- Check how the dough is kneading after five minutes pass because you may need to add either one Tablespoon of water or one Tablespoon of flour based on consistency.
- When the bread is complete, cool it on a wire rack before slicing.
- Enjoy!

NUTRITIONS

- Calories: 180
- Carbs: 33g
- Fat: 2g
- Protein: 7g

Chapter 4

Vegetable Breads

Broccoli and Cauliflower Bread

Broccoli and Cauliflower Bread is for those of you who have a bread machine and want a bread recipe that is tasty and healthy.

Preparation:

2 Hours 20 Minutes

Cooking Time:

50 Minutes

Servings:

1 Loaf

INGREDIENTS

- ¼ cup water
- 4 Tablespoons olive oil
- 1 egg white
- 1 teaspoon lemon juice
- 2/3 cup grated cheddar cheese
- 3 Tablespoons green onion
- ½ cup broccoli, chopped
- ½ cup cauliflower, chopped
- ½ teaspoon lemon pepper seasoning
- 2 cups bread flour
- 1 teaspoon bread machine yeast

DIRECTIONS

- Put all of the mixtures into your bread machine, carefully following the instructions of the manufacturer. Set the program of your bread machine to Basic/White Bread and set crust type to Medium.
- Press starts. Wait until the cycle completes. Once the loaf is ready, take the bucket out and let the loaf cool within 5 minutes. Shake the bucket to remove the loaf. Transfer to a cooling rack, slice then serve.

NUTRITIONS

- Calories: 156 Cal
- Fat: 8 g
- Carbohydrates: 17 g
- Protein: 5 g

Potato Bread

Potato Bread is one of the favorite recipes. As they said, potato bread is the most comfortable bread that you can eat.

Preparation:

3 Hours

Cooking Time:

45 Minutes

Servings:

2 Loaves

INGREDIENTS

- 1 3/4 teaspoons active dry yeast
- 2 Tablespoons dry milk
- 1/4 cup instant potato flakes
- 2 Tablespoons sugar
- 4 cups bread flour
- 1 1/4 teaspoons salt
- 2 Tablespoons butter

- 1 3/8 cups water

DIRECTIONS

- Put all the liquid ingredients in the pan. Add all the dry ingredients, except the yeast. Form a shallow hole in the middle of the dry ingredients and place the yeast.
- Secure the pan in the machine and close the lid. Choose the basic setting and your desired color of the crust. Press starts. Allow the bread to cool before slicing.

NUTRITIONS

- Calories: 35 Cal
- Carbohydrate: 19 g
- Fat: 0 g
- Protein: 4 g

Onion Potato Bread

This bread is perfect for breakfast. The onions and potatoes add a lot of flavor and depth to the bread.

Preparation:

1 Hour 20 Minutes

Cooking Time:

45 Minutes

Servings:

2 Loaves

INGREDIENTS

- 2 Tablespoons quick rise yeast
- 4 cups bread flour
- 1 1/2 teaspoons seasoned salt
- 3 Tablespoons sugar
- 2/3 cup baked potatoes, mashed
- 1 1/2 cups onions, minced
- 2 large eggs
- 3 Tablespoons oil

- 3/4 cup hot water, with a temperature of 115 to 125°F (46 to 51°C)

DIRECTIONS

- Put the liquid ingredients in the pan. Add the dry ingredients, except the yeast. Form a shallow well in the middle using your hand and put the yeast.
- Place the pan in the machine, close the lid, and turn it on. Select the express bake 80 settings and start the machine. Once the bread is cooked, leave it on a wire rack for 20 minutes or until cooled.

NUTRITIONS

- Calories: 160 Cal
- Carbohydrate: 44 g
- Fat: 2 g
- Protein: 6 g

Curd Bread

This bread is slightly tart and very good with butter and jam.

Preparation:

4 Hours

Cooking Time:

15 Minutes

Servings:

12

INGREDIENTS

- ¾ cup lukewarm water
- 3 2/3 cups wheat bread machine flour
- ¾ cup cottage cheese
- 2 Tablespoons softened butter
- 2 Tablespoons white sugar
- 1½ teaspoons sea salt
- 1½ Tablespoons sesame seeds
- 2 Tablespoons dried onions
- 1¼ teaspoons bread machine yeast

DIRECTIONS

- Place all the dry and liquid ingredients in the pan and follow the instructions for your bread machine.
- Pay particular attention to measuring the ingredients. Use a measuring cup, measuring spoon, and kitchen scales to do so. Set the baking program to BASIC and the crust type to MEDIUM.
- When the program has ended, take the pan out of the bread machine and cool for 5 minutes. Shake the loaf out of the pan. If necessary, use a spatula.
- Wrap the bread with a kitchen towel and set it aside for an hour. Otherwise, you can cool it on a wire rack.

NUTRITIONS

- Calories: 277 Cal

- Carbohydrate: 48.4 g
- Fat: 4.7g
- Protein: 9.4 g

Potato Rosemary Bread with Honey

This bread is easy to make and is perfect for sandwiches or soup.

Preparation:

3 Hours

Cooking Time:

30 Minutes

Servings:

20

INGREDIENTS

- 4 cups bread flour, sifted
- 1 Tablespoon white sugar
- 1 Tablespoon sunflower oil
- 1½ teaspoons salt
- 1½ cups lukewarm water
- 1 teaspoon active dry yeast
- ½ Tablespoon honey
- 1 cup potatoes, mashed
- 2 teaspoons crushed rosemary

DIRECTIONS

- Prepare all of the ingredients for your bread and measuring means (a cup, a spoon, kitchen scales). Carefully measure the ingredients into the pan, except the potato and rosemary.
- Place all mixtures into the bread bucket in the right order, following the manual for your bread machine. Close the cover.
- Select the program of your bread machine to Bread with Fillings and choose the crust color to Medium. Press Start.
- After the signal, put your mashed potato and rosemary into the dough. Wait until the program completes.
- When done, take the bucket out

and let it cool for 5-10 minutes. Shake the loaf from the pan and let cool for 30 minutes on a cooling rack. Slice, serve and enjoy the taste of fragrant homemade bread.

NUTRITIONS

- Calories: 106 Cal
- Carbohydrate: 21 g
- Fat: 1 g
- Protein: 2.9 g

Chapter 5.

Spice and Herb Breads

Italian Herb Bread

Preparation:

5 minutes

Cooking Time:

3 hours 5 Minutes

Servings:

14 slices

INGREDIENTS

- 2 Tablespoons margarine
- 2 Tablespoons sugar
- 1½ cups water
- 3 Tablespoons powdered milk
- 1½ teaspoons dried marjoram
- 1½ teaspoons dried basil
- 1½ teaspoons salt
- 4 cups bread flour
- 1¼ teaspoons yeast
- 1½ teaspoons dried thyme

DIRECTIONS

- Add each ingredient to the bread machine in the instruction and at the temperature recommended by your bread machine manufacturer.
- Close the lid, choose the basic bread, medium crust setting on your bread machine, then press start.
- If the bread machine has finished baking, remove the bread.
- Put it on a cooling rack.

NUTRITIONS

- Carbs: 20 g
- Fat: 3 g
- Protein: 4 g
- Calories: 120 Cal

Olive Bread

Preparation:

10 minutes

Cooking Time:

3 hours

Servings:

14 slices

INGREDIENTS

- ½ cup brine from the olive jar
- Add warm water (110°F) To make 1½ cup when combined with brine
- 2 Tablespoons olive oil
- 3 cups bread flour
- 1 2/3 cups whole wheat flour
- 1 ½ teaspoons salt
- 2 Tablespoons sugar
- 1 1/2 teaspoons dried leaf basil
- 2 teaspoons active dry yeast
- 2/3 cup finely chopped Kalamata olives

DIRECTIONS

- Add each ingredient except the olives to the bread machine.
- Close the lid, select the wheat, medium crust setting on your bread machine, and press start.
- Add the olives 10 minutes before the last kneading cycle ends.
- When the bread machine has finished baking, get the bread and put it on a cooling rack.

NUTRITIONS

- Carbs: 28 g
- Fat: 1 g
- Protein: 5 g
- Calories: 140 Cal

Cardamom Cranberry Bread

Preparation:

5 minutes

Cooking Time:

3 hours

Servings:

14 slices

INGREDIENTS

- 1¾ cups water
- 2 Tablespoons brown sugar
- 1½ teaspoons salt
- 2 Tablespoons coconut oil
- 4 cups flour
- 2 teaspoons cinnamon
- 2 teaspoons cardamom
- 1 cup dried cranberries
- 2 teaspoons yeast

DIRECTIONS

- Add each ingredient except the dried cranberries to the bread machine in the order and at the temperature recommended by your bread machine manufacturer.
- Close the lid; select the basic bread setting on your bread machine and press start.
- Add the dried cranberries 5 to 10 minutes before the last kneading cycle ends.
- When the bread machine has stopped baking, remove the bread and put it on a cooling rack.

NUTRITIONS

- Carbs: 41 g
- Fat: 3 g
- Protein: 3 g
- Calories: 157 Cal

Chive Bread

Preparation:

10 minutes

Cooking Time:

3 hours

Servings:

14 slices

INGREDIENTS

- 2/3 cup milk (70°F to 80°F)
- 1/4 cup water (70°F to 80°F)
- 1/4 cup sour cream
- 2 Tablespoons butter
- 1 1/2 teaspoons sugar
- 1 1/2 teaspoons salt
- 3 cups bread flour
- 1/8 teaspoon baking soda
- 1/4 cup minced chives
- 2 1/4 teaspoons active dry yeast leaves

NUTRITIONS

DIRECTIONS

- Add each ingredient to the machine in the order and at the temperature endorsed by your bread machine manufacturer.
- Close the lid, pick the basic bread, medium crust setting on your bread machine, and press start.
- When the bread machine has been finished baking, remove the bread and put it on a cooling rack.

- Carbs: 18 g
- Fat: 2 g
- Protein: 4 g
- Calories: 105 Cal

Lavender Buttermilk Bread

Preparation:

10 minutes

Cooking Time:

3 hours

Servings:

14 slices

INGREDIENTS

- ½ cup water
- 7/8 cup buttermilk
- 1/4 cup olive oil
- 3 Tablespoons fresh lavender leaves, finely chopped
- 1 1/4 tsp. finely chopped fresh lavender flowers
- 1 lemon, Grated
- 4 cups bread flour
- 2 teaspoons salt
- 2 3/4 teaspoons bread machine yeast

DIRECTIONS

- Add each ingredient to the bread machine in the order and at the temperature endorsed by your bread machine manufacturer.
- Close the lid, select the quick, medium crust setting on your bread machine, and press start.
- When the bread machine has finished baking, take out the bread and put it on a cooling rack.

NUTRITIONS

- Carbs: 27 g
- Fat: 5 g
- Protein: 2 g
- Calories: 160 Cal

Chapter 6.

Cheese Bread

Cream Cheese Rolls

Cream Cheese Rolls is a cooked bread roll of the type often eaten for breakfast. They are slightly sweet and delicious. Everyone should know how to make them.

Preparation:

10 minutes

Cooking Time:

40 Minutes

Servings:

6

INGREDIENTS

- 3 eggs
- 3 oz. Full-fat cream cheese, cubed and cold
- ¼ teaspoon Cream of tartar
- ¼ teaspoon salt

DIRECTIONS

- Preheat the oven to 300°F. Line a baking pan with parchment paper. Grease with cooking oil. Remove and separate the yolks from the eggs and place the whites in a container. Whisk with the tartar until stiff.
- In another container, whisk the cream cheese, salt, and yolks until smooth.
- Fold in the whites of the eggs, mixing well with a spatula. Mold a scoop of whites over the yolk mixture and fold together as you rotate the dish. Continue unit well combined.
- Place six large spoons of the mixture onto the prepared pan.

Mash the tops with the spatula to flatten slightly.

- Bake until browned, about 30 to 40 minutes. Cool a few minutes in the pan. Then, carefully arrange them on a wire rack to cool.

NUTRITIONS

- Calories: 91.3 Cal
- Fat: 8 g
- Carb: 1 g
- Protein: 4.2 g

Cheese Blend Bread

Cheese Blend Bread is an excellent way to use up pieces of cheeses leftover at the end of the week. It is a perfect bread, and there is nothing else quite like it. This recipe will make enough for a large family.

Preparation:

45 minutes

Cooking Time:

20 Minutes

Servings:

12

INGREDIENTS

- 5 oz. cream cheese
- 1/4 cup ghee
- 2/3 cup almond flour
- 1/4 cup coconut flour
- 3 Tablespoons whey protein, unflavored
- 2 teaspoon baking powder
- 1/2 teaspoon Himalayan salt
- 1/2 cup parmesan cheese, shredded
- 3 Tablespoons water
- 3 eggs

- 1/2 cup mozzarella cheese, shredded

DIRECTIONS

- Place wet ingredients into bread machine pan. Add dry ingredients. Set the bread machine to the gluten-free setting.
- When the bread is done, remove the bread machine pan from the bread machine. Let cool slightly before transferring to a cooling rack. You can store your bread for up to 5 days.

NUTRITIONS

- Calories: 132 Cal
- Carbohydrates: 4 g
- Protein: 6 g
- Fat: 8 g

Cheese Spinach Crackers

The plain taste gets into your head and stays there for a while. It can be served as a starter or as a snack. The cheese spinach crackers are very popular among children.

Preparation:

15 minutes

Cooking Time:

25 Minutes

Servings:

16

INGREDIENTS

- 1 ½ cups almond flour
- 150 g fresh spinach
- ½ cup flax meal
- ¼ cup coconut flour
- ½ teaspoon ground cumin
- ¼ cup butter
- ½ cup parmesan cheese, grated
- ½ teaspoon flaked chili peppers, dried
- ½ teaspoon salt

DIRECTIONS

- Bring water to boil in a saucepan. Add spinach and cook for 1 minute. Add cooked spinach leaves into a cold-water bowl to stop the cooking process.
- Squeeze out the water from the spinach leaves and drain. Process the spinach in a food processor, process until a smooth consistency is reached.
- In the meantime, add almond flour, coconut flour, flax meal, cumin, chili flakes, salt, and parmesan cheese into the bowl and mix well.
- Add softened butter and spinach into the flour mixture

and mix to combine well—transfer dough into a refrigerator. Wrap in foil and keep for 1 hour.

- Preheat oven to 400°F. Remove the foil wrapping and transfer the dough to a parchment paper-lined baking sheet.
- Top dough with second parchment paper piece and roll dough with a rolling pin until the dough is ¼ inch thick.
- Slice dough into 16 even pieces using a pizza cutter. Move the baking sheet into the warm oven and bake the dough for 18 to 20 minutes.
- For a crunchier texture, adjust oven temperature to 260°F and bake for 15 to 20 minutes more.

NUTRITIONS

- Calories: 126 Cal
- Fat: 10.9 g
- Carb: 1.4 g
- Protein: 4.5 g

Ricotta Bread

This bread is healthier than the typical buttered bread because it's made from bread flour instead of wheat flour. It's delicious, easy to make, and goes great with any meal!

Preparation:

3 Hours

Cooking Time:

30 Minutes

Servings:

10

INGREDIENTS	DIRECTIONS
• 1/3 cup milk • 1 cup ricotta cheese • 2 Tablespoons butter • 1 egg • 2 ½ Tablespoons sugar • 1 teaspoon salt • 2 ¼ cups bread flour • 1 ½ teaspoons yeast	• Put all bread mixtures in your bread machine, in the order listed above, starting with the milk, and finishing with the yeast. • Create or make a well in the center of your flour and place the yeast in the well. Make sure the well doesn't touch any liquid. Set the bread machine to the basic function with a light crust. • Check on the dough after about 5 minutes and make sure that it's a softball. Put water 1

tablespoon at a time if it's too dry, and add flour 1 tablespoon if it's too wet. When the bread is done, allow it cool on a wire rack.

NUTRITIONS

- Calories: 115 Cal
- Fat: 6.5 g
- Carbs: 3.3 g
- Protein: 8.5 g

Bacon Jalapeño Cheesy Bread

This bread is heaty! It's got an excellent salty heat from the bacon, the pepper's bite, and the savory of the cheese.

Preparation:

5 minutes

Cooking Time:

40 Minutes

Servings:

12

INGREDIENTS

- 1 cup golden flaxseed, ground
- 3/4 cup coconut flour
- 2 teaspoons baking powder
- 1/4 teaspoon black pepper
- 1 Tablespoon erythritol
- 1/3 cup pickled jalapeno
- 8 oz. cream cheese, full fat
- 4 eggs
- 3 cups shredded sharp cheddar
- cheese, + 1/4 cup extra for the
- topping
- 3 Tablespoon parmesan cheese,
- grated
- 1 1/4 cups almond milk
- 5 bacon slices (cooked and
- crumbled)
- 1/4 cup rendered bacon grease

  (from frying the bacon)

DIRECTIONS

- Cook the bacon in a larger frying pan, set aside to cool on paper towels. Save 1/4 cup of bacon fat for the recipe; let cool slightly before using.
- 2.   Add wet ingredients to the bread machine pan, including the cooled bacon grease. Add in the remaining ingredients.
- 3.   When the bread is done, set to a quick bread setting, remove the bread machine pan from the bread machine.
- 4.   Let cool slightly before transferring to a cooling rack. Once on a cooling rack, top with the remaining cheddar cheese. You can store your bread for up to 7 days.

NUTRITIONS

- Calories: 235 Cal
- Carbohydrates: 5 g
- Protein: 11 g
- Fat: 17 g

Chapter 7.

Fruit Bread

Spicy Fruit Bread

Preparation:

1 hour 40 minutes

Cooking Time:

40-45 minutes

Servings:

1 Loaf

INGREDIENTS

- 3½ teaspoon dry yeast
- 300 g wholemeal flour
- 100 g white flour
- 1 teaspoon salt
- 75 g butter
- 2 teaspoon sugar
- 2 teaspoon cinnamon
- 1 teaspoon mixed spice
- 2 eggs medium
- 110 ml milk
- 110 ml water

- 150 g mixed dried fruit

DIRECTIONS

- Place the ingredients (except dried fruit) in the bread pan in the order listed in the recipe above.
- Place mixed dried fruit into automatic raisin and nut dispenser.
- Select Menu '06' (5hr), make sure the size is set to M, and press start.
- When the cycle is complete, turn out and let cool before slicing/serving.

NUTRITIONS

- Calories: 310 Cal
- Total Carbohydrate: 40 g
- Fat: 13 g
- Protein: 3 g

Orange and Walnut Bread

Preparation:

2 hour 50 minutes

Cooking Time:

45 minutes

Servings:

10-15

INGREDIENTS

- 1 egg white
- 1 Tablespoon water
- ½ cup warm whey
- 1 Tablespoon yeast
- 4 Tablespoons sugar
- 2 oranges, crushed
- 4 cups flour
- 2 ½ Tablespoons salt
- 3 teaspoons orange peel
- 1/3 teaspoon vanilla
- 3 Tablespoons walnut and almonds, crushed

- Crushed pepper, cheese for garnish

DIRECTIONS

- Add all of the ingredients to your Bread Machine (except egg white, 1 Tablespoon water and crushed pepper/ cheese).
- Set the program to "Dough" cycle and let the cycle run.
- Remove the dough (using lightly floured hands) and carefully place it on a floured surface.
- Cover with a light film/cling paper and let the dough rise for 10 minutes.
- Divide the dough into thirds after it has risen
- Place on a light flour surface, roll each portion into 14x10 inch sized rectangles
- Use a sharp knife to cut carefully, cut the dough into strips of ½ inch width

- Pick 2-3 strips and twist them multiple times, making sure to press the ends together
- Preheat your oven to 400°F
- Take a bowl and stir egg white, water, and brush onto the breadsticks
- Sprinkle salt, pepper/ cheese
- Bake for 10-12 minutes until golden brown
- Remove from baking sheet and transfer to a cooling rack. Serve and enjoy!

NUTRITIONS

- Calories: 437 Cal
- Total Carbohydrate: 82 g
- Total Fat: 7 g
- Protein: 12 g
- Sugar: 34 g
- Fiber: 1 g

Apple with Pumpkin Bread

Preparation:

2 hour 50 minutes

Cooking Time:

45 minutes

Servings:

2 Loaf

INGREDIENTS

- 1/3 cup dried apples, chopped
- 1 1/2 teaspoon bread machine yeast
- 4 cups bread flour
- 1/3 cup ground pecans
- 1/4 teaspoon ground nutmeg
- 1/4 teaspoon ground ginger
- 1/4 teaspoon allspice
- 1/2 teaspoon ground cinnamon
- 1 1/4 teaspoon salt
- 2 Tablespoons unsalted butter, cubed
- 1/3 cup dry skim milk powder
- 1/4 cup honey
- 2 large eggs, at room temperature
- 2/3 cup pumpkin puree

- 2/3 cup water, with a temperature of 80 to 90°F (26 to 32°C)

DIRECTIONS

- Put all ingredients, except the dried apples, in the bread pan in this order: water, pumpkin puree, eggs, honey, skim milk, butter, salt, allspice, cinnamon, pecans, nutmeg, ginger, flour, and yeast.
- Secure the pan in the machine and lock the lid.
- Place the dried apples in the fruit and nut dispenser.
- Turn on the machine. Choose the sweet setting and your desired color of the crust.
- Carefully unmold the baked bread once done and let cool for 20 minutes before slicing.

NUTRITIONS

- Calories: 228 Cal
- Total Carbohydrate: 30 g
- Total Fat: 4 g
- Protein: 18 g

Warm Spiced Pumpkin Bread

Preparation:

2 hour

Cooking Time:

15 minutes

Servings:

12-16

INGREDIENTS

- Butter for greasing the bucket
- 1 1/2 cups pumpkin purée
- 3 eggs, at room temperature
- 1/3 cup melted butter cooled
- 1 cup sugar
- 3 cups all-purpose flour
- 1 1/2 teaspoons baking powder
- ¾ teaspoon ground cinnamon
- ½ teaspoon baking soda
- ¼ teaspoon ground nutmeg
- ¼ teaspoon ground ginger
- ¼ teaspoon salt
- Pinch ground cloves

DIRECTIONS

- Lightly grease the bread bucket with butter.
- Add the pumpkin, eggs, butter, and sugar.
- Program the machine for Quick/Rapid bread and press Start.
- Let the wet ingredients be mixed by the paddles until the first fast mixing cycle is finished, about 10 minutes into the cycle.
- Stir together the flour, baking powder, cinnamon, baking soda, nutmeg, ginger, salt, and cloves until well blended.
- Add the dry ingredients to the bucket when the second fast mixing cycle starts.
- When the loaf is done, remove the bucket from the machine.

- Let the loaf cool for 5 minutes.
- Gently shake the bucket to remove the loaf and turn it out onto a rack to cool..

NUTRITIONS

- Calories: 251 Cal
- Total Carbohydrate: 43 g
- Total Fat: 7 g
- Protein: 5 g
- Sodium: 159 mg
- Fiber: 2 g

Date Delight Bread

Preparation:

2 hour

Cooking Time:

15 minutes

Servings:

12

INGREDIENTS

- ¾ cup water, lukewarm
- ½ cup milk, lukewarm
- 2 tablespoons butter, melted at room temperature
- ¼ cup honey
- 3 tablespoons molasses
- 1 tablespoon sugar
- 2 ¼ cups whole-wheat flour
- 1 ¼ cups white bread flour
- 2 tablespoons skim milk powder
- 1 teaspoon salt
- 1 tablespoon unsweetened cocoa powder
- 1 ½ teaspoons instant or bread machine yeast

- ¾ cup chopped dates

DIRECTIONS

- Take 1 ½ pound size loaf pan and add the liquid ingredients and then add the dry ingredients. (Do not add the dates as of now.)
- Place the loaf pan in the machine and close its top lid.
- Plug the bread machine into the power socket. For selecting a bread cycle, press "Basic Bread/White Bread/Regular Bread" or "Fruit/Nut Bread," and for selecting a crust type, press "Light" or "Medium".
- Start the machine, and it will start preparing the bread. When the machine beeps or signals, add the dates.
- After the bread loaf is completed, open the lid and take out the loaf pan.

- Allow the pan to cool down for 10-15 minutes on a wire rack. Gently shake the pan and remove the bread loaf.
- Make slices and serve.

NUTRITIONS

- Calories: 220 Cal
- Total Carbohydrate: 52 g
- Cholesterol: 0 g
- Total Fat: 5 g
- Protein: 4 g

Chapter 8.

Meat Bread

French Ham Bread

Preparation:

30-45 Minutes

Cooking Time:

2 Hours

Servings:

8

INGREDIENTS

- 3 1/3 cups wheat flour
- 1 cup ham
- ½ cup of milk powder
- 1 ½ Tablespoons sugar
- 1 teaspoon yeast, fresh
- 1 teaspoon salt
- 1 teaspoon dried basil
- 1 1/3 cups water
- 2 Tablespoons olive oil

DIRECTIONS

- Cut ham into cubes of 0.5-1 cm (approximately ¼ inch).
- Put the ingredients in the bread maker in the following order: water, olive oil, salt, sugar, flour, milk powder, ham, and yeast.
- Put all the ingredients according to the instructions in your bread maker.
- Basil put in a dispenser or fill it later, at the signal in the container.
- Turn on the bread maker.
- After the baking cycle, leave the bread container in the bread maker to keep warm for 1 hour.
- Then your delicious bread is ready!

NUTRITIONS

- Calories 287 Cal
- Total Fat 5.5 g
- Saturated Fat 1.1 g
- Cholesterol 11 g
- Sodium 557 mg
- Total Carbohydrate 47.2 g
- Dietary Fiber 1.7 g
- Total Sugars 6.4 g
- Protein 11.4 g

Onion Bacon Bread

Preparation:

1 hour 30 Minutes

Cooking Time:

1 Hour 30 minutes

Servings:

8

INGREDIENTS

- 1 ½ cups water
- 2 Tablespoons sugar
- 3 teaspoons dry yeast
- 4 ½ cups flour
- 1 egg
- 2 teaspoons salt
- 1 Tablespoon oil
- 3 small onions, chopped

- 1 cup bacon

NUTRITIONS

DIRECTIONS

- Cut the bacon.
- Put all the ingredients into the bread machine.
- Set it to the Basic program.
- Enjoy this tasty bread!

- Calories: 391 Cal
- Total Fat: 9.7 g
- Saturated Fat: 2.7 g
- Cholesterol: 38 g
- Sodium: 960 mg,
- Total Carbohydrate: 59.9 g
- Dietary Fiber: 2.8 g
- Total Sugars: 4.3 g
- Protein: 14.7 g

Sausage Bread

Preparation:

2 hours

Cooking Time:

2 Hours

Servings:

8

INGREDIENTS

- 1 ½ teaspoons dry yeast
- 3 cups flour
- 1 teaspoon sugar
- 1 ½ teaspoons salt
- 1 1/3 cups whey
- 1 Tablespoon oil
- 1 cup smoked sausage, chopped

NUTRITIONS

DIRECTIONS

- Fold all the ingredients in the order that is recommended specifically for your model.
- Set the required parameters for baking bread.
- When ready, remove the delicious hot bread.
- Wait for it to cool down and enjoy it with sausage.

- Calories: 234 Cal
- Total Fat: 5.1 g
- Saturated Fat: 1.2 g
- Cholesterol: 9 g
- Sodium: 535 mg
- Total Carbohydrate: 38.7 g
- Dietary Fiber: 1.4 g
- Total Sugars: 2.7 g
- Protein: 7.4 g

Crazy Crust Pizza Dough

Preparation:

10 Minutes

Cooking Time:

45 Minutes

Servings:

8

INGREDIENTS

- 1 cup all-purpose flour
- 1 teaspoon salt
- 1 teaspoon dried oregano
- 1/8 teaspoon black pepper
- 2 eggs, lightly beaten
- 2/3 cup milk

DIRECTIONS

- Heat oven to 200°C or 400°F. Grease a baking sheet or rimmed pizza pan lightly.
- Mix black pepper, oregano, salt, and flour in a big bowl. Stir in milk and eggs thoroughly. Put the batter in the pan and tilt it until it is evenly coated. Put whatever toppings you want on top of the batter.
- Bake it in the oven until the crust is set for 20-25 minutes.
- Take the crust out of the oven. Drizzle pizza sauce on and top with cheese. Bake for around 10 minutes until the cheese melts.

NUTRITIONS

- Calories: 86 Cal
- Total Carbohydrate: 13.1 g
- Cholesterol: 48 mg
- Total Fat: 1.8 g
- Protein: 3.9 g
- Sodium: 317 mg

Double Crust Stuffed Pizza

Preparation:

30 Minutes

Cooking Time:

2 Hours 45 Minutes

Servings:

8

INGREDIENTS

- 1 1/2 teaspoons white sugar
- 1 cup warm water (100°F/40°C)
- 1 1/2 teaspoons active dry yeast
- 1 Tablespoon olive oil
- 1/2 teaspoon salt
- 2 cups all-purpose flour
- 1 (8 oz.) can crushed tomatoes
- 1 Tablespoon packed brown sugar
- 1/2 teaspoon garlic powder
- 1 teaspoon olive oil
- 1/2 teaspoon salt
- 3 cups shredded mozzarella cheese, divided
- 1/2 lb. bulk Italian sausage

DIRECTIONS

- In a large bowl or work bowl of a stand mixer, mix warm water and white sugar. Sprinkle with yeast and let the mixture stand for 5 minutes until the yeast starts to form creamy foam and softens. Stir in 1 Tablespoon of olive oil.
- Mix flour with 1/2 teaspoon of salt. Add half of the flour mixture into the yeast mixture and mix until no dry spots are visible. Whisk in remaining flour, half cup at a time, mixing well every after addition. Place the dough on a lightly floured surface once it has pulled together. Knead the dough for 8 minutes until elastic and smooth. You can use the dough hook in a stand mixer to mix the dough.
- Transfer the dough into a lightly oiled large bowl and flip to coat the dough with oil. Use a light cloth to cover the dough. Let it rise in a warm place for 1 hour until the volume doubles.
- In a small saucepan, mix 1 teaspoon of olive oil, brown sugar, crushed tomatoes, garlic powder, and salt. Cover the saucepan and let it cook over low

- 1 (4 oz.) package sliced pepperoni
- 1 (8 oz.) package sliced fresh mushrooms
- 1/2 green bell pepper, chopped
- 1/2 red bell pepper, chopped

heat for 30 minutes until the tomatoes begin to break down.

- Set the oven to 450°F (230°C) for preheating. Flatten the dough and place it on a lightly floured surface. Divide the dough into 2 equal portions. Roll one portion into a 12-inches thin circle. Roll the other portion into a 9-inches thicker circle.
- Press the 12-inches dough round into an ungreased 9-inches springform pan. Top the dough with a cup of cheese. Form sausage into a 9-inches patty and place it on top of the cheese. Arrange pepperoni, green pepper, mushrooms, red pepper, and remaining cheese on top of the sausage patty. Place the 9-inches dough round on top, pinching its edges to seal. Make vent holes on top of the crust by cutting several 1/2-inch on top. Pour the sauce evenly on top of the crust, leaving an only 1/2-inch border at the edges.
- Bake the pizza inside the preheated oven for 40-45 minutes until the cheese is melted, the sausage is cooked through, and the crust is fixed. Let the pizza rest for 15 minutes. Before serving, cut the pizza into wedges.

NUTRITIONS

- Calories: 410 Cal
- Total Carbohydrate: 32.5 g
- Cholesterol: 53 mg
- Total Fat: 21.1 g
- Protein: 22.2 g
- Sodium: 1063 mg

Chapter 9.

Sweet Breads

Sweet Almond Anise Bread

Preparation:

2 Hours 20 Minutes

Cooking Time:

50 Minutes

Servings:

1 Loaf

INGREDIENTS

- ¾ cup water
- ¼ cup butter
- ¼ cup sugar
- ½ teaspoon salt
- 3 cups bread flour
- 1 teaspoon anise seed
- 2 teaspoons active dry yeast
- ½ cup almonds, chopped

DIRECTIONS

- Add all of the ingredients to your bread machine, carefully following the instructions of the manufacturer
- Set the program of your bread machine to Basic/White Bread and set crust type to Medium
- Press START
- Wait until the cycle completes
- Once the loaf is ready, take the bucket out and let the loaf cool for 5 minutes
- Gently shake the bucket to remove the loaf
- Transfer to a cooling rack, slice, and serve
- Enjoy!

NUTRITIONS

- Calories: 87 Cal
- Fat: 4 g
- Carbohydrates: 7 g
- Protein: 3 g
- Fiber: 1 g

Chocolate Chip Peanut Butter Banana Bread

Preparation:

25 Minutes

Cooking Time:

10 Minutes

Servings:

10-16 Slices

INGREDIENTS

- Two bananas, mashed
- Two eggs, at room temperature
- 1/2 cup melted butter, cooled
- Two tablespoons milk, at room temperature
- One teaspoon pure vanilla extract
- 2 cups all-purpose flour
- 1/2 cup sugar
- 11/4 teaspoons baking powder
- 1/2 teaspoon baking soda
- 1/2 teaspoon salt
- 1/2 cup peanut butter chips
- 1/2 cup semisweet chocolate chips

DIRECTIONS

- Stir together the bananas, eggs, butter, milk, and vanilla in the bread machine bucket and set it aside.
- In a medium bowl, toss together the flour, sugar, baking powder, baking soda, salt, peanut butter chips, and chocolate chips.
- Add the dry ingredients to the bucket.
- Program the machine for Quick/Rapid bread, and press Start.
- When the cake is made, stick a knife into it, and if it arises out clean, the loaf is done.
- If the loaf needs a few more minutes, look at the management panel for a Bake Only button, and extend the time by 10 minutes.
- When the loaf is done, remove the bucket from the machine.
- Let the loaf cool for 5 minutes.
- Gently rock the can to remove the bread and turn it out onto a rack to cool.

NUTRITIONS

- Calories: 297
- Total Fat: 14g
- Saturated Fat: 7g
- Carbohydrates: 40g
- Fiber: 1g
- Sodium: 255mg
- Protein: 4g

Nectarine Cobbler Bread

Preparation:

10 Minutes

Cooking Time:

5 Minutes

Servings:

12-16 Slices

INGREDIENTS

- 1/2 cup (1 stick) butter, at room temperature
- Two eggs, at room temperature
- 1 cup of sugar
- 1/4 cup milk, at room temperature
- One teaspoon pure vanilla extract
- 1 cup diced nectarines
- 1 3/4 cups all-purpose flour
- One teaspoon baking soda
- 1/2 teaspoon salt
- 1/2 teaspoon ground nutmeg
- 1/4 teaspoon baking powder

DIRECTIONS

- Place the butter, eggs, sugar, milk, vanilla, and nectarines in your bread machine.
- Program the machine for Quick/Rapid bread and press Start.
- While the wet ingredients are mixing, stir together the flour, baking soda, salt, nutmeg, and baking powder in a small bowl.
- After the first fast mixing is done and the machine signals, add the dry ingredients.
- When the loaf is done, remove the bucket from the machine.
- Let the loaf cool for 5 minutes.
- Gently shake the bucket to remove the loaf, then turn it out onto a rack to cool.

NUTRITIONS

- Calories: 218
- Total Fat: 9g
- Saturated Fat: 5g
- Carbohydrates: 32g
- Fiber: 1g
- Sodium: 270mg
- Protein: 3g

Hot Buttered Rum Bread

Preparation:

10 Minutes

Cooking Time:

3 Hours 40 Minutes

Servings:

1 Loaf

INGREDIENTS

- 1 egg
- 1 Tablespoon rum extract
- 3 Tablespoons butter, softened
- 3 cups bread flour
- 3 Tablespoons packed brown sugar
- 1 ¼ teaspoons salt
- 1/2 teaspoon ground cinnamon
- 1/4 teaspoon ground nutmeg
- 1/4 teaspoon ground cardamom
- 1 teaspoon bread machine or quick active dry yeast

Topping:

- 1 egg yolk, beaten
- 1 ½ teaspoons finely chopped pecans
- 1 ½ teaspoons packed brown sugar

DIRECTIONS

- Directions:
- Break the egg into 1 cup, and add water to fill out the measuring cup
- Place egg mixture and bread ingredients into a pan.
- Choose basic bread setting and medium/light crust color.
- While bread bakes, combine topping ingredients in a small bowl and brush on top of bread when 40 – 50 minutes are remaining of the cooking time.

NUTRITIONS

- Calories: 170 Cal
- Total Carbohydrate: 31 g
- Cholesterol: 25 mg
- Total Fat: 2.0 g
- Protein: 4 g
- Sodium: 270 mg
- Fiber: 1 g

Almond and Chocolate Chip Bread

Preparation:

10 Minutes

Cooking Time:

3 Hours 40 Minutes

Servings:

1 Loaf

INGREDIENTS

- 1 cup plus 2 Tablespoons water
- 2 Tablespoons butter or margarine, softened
- ½ teaspoon vanilla
- 3 cups Gold Medal™ Better for Bread™ flour
- ¾ cup semisweet chocolate chips
- 3 Tablespoons sugar
- 1 Tablespoon dry milk
- ¾ teaspoon salt
- 1 ½ teaspoons bread machine or quick active dry yeast
- 1/3 cup sliced almonds

DIRECTIONS

- Measure and put all ingredients except almonds in the bread machine pan. Add almonds at the Nut signal or 5 - 10 minutes before the kneading cycle ends.
- Select the white cycle. Use Light crust color.
- Take out baked bread from the pan.

NUTRITIONS

- Calories: 130 Cal
- Total Carbohydrate: 18 g
- Total Fat: 7 g
- Protein: 1 g
- Protein: 3 g

Chapter 10

Sourdough Breads

Garlic and Herb Flatbread Sourdough

Preparation:

1 Hour

Cooking Time:

25-30 Minutes

Servings:

12

INGREDIENTS	DIRECTIONS
Dough: • 1 cup sourdough starter, fed or unfed • 3/4 cup warm water • 2 teaspoons instant yeast • 3 cups all-purpose flour • 1 1/2 teaspoons salt • 3 Tablespoons olive oil Topping: • 1/2 teaspoon dried thyme • 1/2 teaspoon dried oregano • 1/2 teaspoon dried marjoram • 1 teaspoon garlic powder • 1/4 teaspoon onion powder • 1/4 teaspoon salt • 1/4 teaspoon pepper • 3 Tablespoons olive oil	• Combine all the dough ingredients in the bowl of a stand mixer, and knead until smooth. Place in a lightly greased bowl and let rise for at least one hour. Punch down, then let rise again for at least one hour. • To prepare the topping, mix all ingredients except the olive oil in a small bowl. • Lightly grease a 9x13 baking pan or standard baking sheet, and pat and roll the dough into a long rectangle in the pan. Brush the olive oil over the dough, and sprinkle the herb and seasoning mixture over top. Cover and let rise for 15-20 minutes. • Preheat oven to 425°F and bake for 25-30 minutes.

NUTRITIONS

- Calories: 89 Cal
- Fat: 3.7 g
- Protein: 1.8 g

Sauerkraut Rye

Preparation:

2 Hours 20 Minutes

Cooking Time:

50 Minutes

Servings:

1 Loaf

INGREDIENTS

- 1 cup sauerkraut, rinsed and drained
- ¾ cup warm water
- 1½ Tablespoons molasses
- 1½ Tablespoons butter
- 1½ Tablespoons brown sugar
- 1 teaspoon caraway seeds
- 1½ teaspoons salt
- 1 cup rye flour
- 2 cups bread flour
- 1½ teaspoons active dry yeast

DIRECTIONS

- Add all of the ingredients to your bread machine.
- Set the program of your bread machine to Basic/White Bread and set crust type to Medium
- Press START
- Wait until the cycle completes
- Once the loaf is ready, take the bucket out and let the loaf cool for 5 minutes
- Gently shake the bucket to remove the loaf
- Transfer to a cooling rack, slice, and serve

NUTRITIONS

- Calories: 74 Cal
- Fat: 2 g
- Carbohydrates: 12 g
- Protein: 2 g
- Fiber: 1 g

Honey Sourdough Bread

Preparation:

15 minutes; 1 week
(Starter)

Cooking Time:

3 Hours

Servings:

1 Loaf

INGREDIENTS	DIRECTIONS
<ul><li>2/3 cup sourdough starter</li><li>1/2 cup water</li><li>1 Tablespoon vegetable oil</li><li>2 Tablespoons honey</li><li>1/2 teaspoon salt</li><li>1/2 cup high protein wheat flour</li><li>2 cups bread flour</li><li>1 teaspoon active dry yeast</li></ul>	<ul><li>Measure 1 cup of starter and remaining bread ingredients, add to bread machine pan.</li><li>Choose basic/white bread cycle with medium or light crust color.</li></ul>

NUTRITIONS

- Calories: 175 Cal
- Total Carbohydrate: 33 g
- Total Fat: 0.3 g
- Protein: 5.6 g
- Sodium: 121 mg
- Fiber: 1.9 g

Olive and Garlic Sourdough Bread

Preparation:

15 minutes; 1 week
(Starter)

Cooking Time:

3 Hours

Servings:

1 Loaf

INGREDIENTS

- 2 cups sourdough starter
- 3 cups flour
- 2 Tablespoons olive oil
- 2 Tablespoons sugar
- 2 teaspoons salt
- 1/2 cup black olives, chopped
- 6 cloves garlic, chopped

DIRECTIONS

- Add starter and bread ingredients to the bread machine pan.
- Choose the dough cycle.
- Conventional Oven:
- Preheat oven to 350°F.
- When the cycle is complete, if the dough is sticky add more flour.
- Shape dough onto a baking sheet or put into loaf pan
- Bake for 35- 45 minutes until golden.
- Cool before slicing.

NUTRITIONS

- Calories: 150 Cal
- Total Carbohydrate: 26.5 g
- Total Fat: 0.5 g
- Protein: 3.4 g
- Sodium: 267 mg
- Fiber: 1.1 g

French Sourdough Bread

Preparation:

15 minutes; 1 week (Starter)

Cooking Time:

3 Hours

Servings:

2 Loaf

INGREDIENTS

- 2 cups sourdough starter
- 1 teaspoon salt
- 1/2 cup water
- 4 cups white bread flour
- 2 Tablespoons white cornmeal

DIRECTIONS

- Add ingredients to bread machine pan, saving cornmeal for later.
- Choose dough cycle.
- Conventional Oven:
- Preheat oven to 375°F.
- At end of dough cycle, turn dough out onto a floured surface.
- Add flour if dough is sticky.
- Divide dough into 2 portions and flatten into an oval shape 1 ½ inch thick.
- Fold ovals in half lengthwise and pinch seams to elongate.
- Sprinkle cornmeal onto the baking sheet and place the loaves seam side down.
- Cover and let rise in until about doubled.
- Place a shallow pan of hot water on the lower shelf of the oven;
- Use a knife to make shallow, diagonal slashes in tops of loaves

- Place the loaves in the oven and spray with a fine water mister. Spray the oven walls as well.
- Repeat spraying 3 times at one-minute intervals.
- Remove pan from water after 15 minutes of baking
- Fully bake for 30 to 40 minutes or until golden brown

NUTRITIONS

- Calories: 937 Cal
- Total Carbohydrate: 196 g
- Total Fat: 0.4 g
- Protein: 26.5 g
- Sodium: 1172 mg
- Fiber: 7.3 g

Chapter 11.

Basic Breads

Basic White Bread

Preparation:

1 hour 15 minutes

Cooking Time:

50 minutes (20+30 minutes)

Servings:

1 Loaf

INGREDIENTS

- ½ to 5/8 cup Water
- 5/8 cup Milk
- 1 ½ Tablespoons butter or margarine
- 3 Tablespoons Sugar
- 1 ½ teaspoons Salt
- 3 cups Bread Flour
- 1 ½ teaspoons Active Dry Yeast

DIRECTIONS

- Put all ingredients in the bread pan, using a minimal measure of liquid listed in the recipe.
- Select medium Crust setting and press Start.
- Observe the dough as it kneads. Following 5 to 10 minutes, in the event that it seems dry and firm, or if your machine seems as though it's straining to knead, add more liquid, 1 tablespoon at a time until dough forms well.
- Once the baking cycle ends, remove bread from pan, and let cool before slicing.

NUTRITIONS

Calories: 64 Cal

Fat: 1 g

Carbohydrates: 12 g

Protein: 2 g

All-Purpose White Bread

Preparation:

2 hours 10 minutes

Cooking Time:

40 Minutes

Servings:

1 Loaf

INGREDIENTS

- ¾ cup water at 80°F
- 1 Tablespoon melted butter, cooled
- 1 Tablespoon sugar
- ¾ teaspoon salt
- 2 Tablespoons skim milk powder
- 2 cups white bread flour
- ¾ teaspoon instant yeast

DIRECTIONS

- Add all of the ingredients to your bread machine, carefully following the instructions of the manufacturer.
- Set the program of your bread machine to Basic/White Bread and set crust type to Medium.
- Press START.
- Wait until the cycle completes.
- Once the loaf is ready, take the bucket out and let the loaf cool for 5 minutes.
- Gently shake the bucket to remove the loaf.
- Put to a cooling rack, slice, and serve.

NUTRITIONS

Calories: 140 Cal

- Fat: 2 g
- Carbohydrates: 27 g
- Protein: 44 g
- Fiber: 2 g

Country White Bread

Preparation:

3 hours

Cooking Time:

45 Minutes

Servings:

2 Loaves

INGREDIENTS

- 2 teaspoon active dry yeast
- 1 1/2 Tablespoons sugar
- 4 cups bread flour
- 1 1/2 teaspoons salt
- 1 large egg
- 1 1/2 Tablespoons butter
- 1 cup warm milk, with a temperature of 110 to 115°F (43 to 46°C)

DIRECTIONS

- Put all the liquid ingredients in the pan. Add all the dry ingredients, except the yeast. Use your hand to form a hole in the middle of the dry ingredients. Put the yeast in the hole.
- 2.   Secure the pan in the chamber and close the lid. Choose the basic setting and your preferred crust color. Press start.
- 3.   Once done, transfer the baked bread to a wire rack. Slice once cooled.

NUTRITIONS

- ○ Calories: 105 Cal
- ○ Total Carbohydrate: 0 g
- ○ Total Fat: 0 g
- ○ Protein: 0 g

Anadama Bread

Preparation:

3 hours

Cooking Time:

45 Minutes

Servings:

2 Loaves

INGREDIENTS

- 3 teaspoon bread machine yeast
- 4 teaspoon vital wheat gluten
- 4 cups bread flour
- 1 teaspoon salt
- 1 cup instant or regular oatmeal
- 2 Tablespoons maple syrup
- 2 Tablespoons unsalted butter, cubed
- 1/3 cup water, with a temperature of 80 to 90°F (26 to 32°C)
- 1 1/2 cups buttermilk, with a temperature of 80 to 90°F (26 to 32°C)

DIRECTIONS

- Put the ingredients in the pan in this order: buttermilk, water, butter, maple syrup, oatmeal, salt, flour, gluten, and yeast.
- Secure the pan in the machine, close the lid, and turn it on.
- Choose the basic setting and your preferred crust color and press start.
- Transfer the baked bread to a wire rack and let cool before slicing.

NUTRITIONS

- Calories: 269 Cal
- Total Carbohydrate: 49 g
- Total Fat: 4 g
- Protein: 8 g

Buttermilk White Bread

Preparation:

2 hour 50 minutes

Cooking Time:

25 Minutes

Servings:

1 Loaf

INGREDIENTS

- 1 1/8 cups water
- 3 teaspoons honey
- 1 Tablespoon margarine
- 1 1/2 teaspoons salt
- 3 cups bread flour
- 2 teaspoons active dry yeast
- 4 teaspoons powdered buttermilk

DIRECTIONS

- Into the pan of the bread machine, place the ingredients in the order suggested by the manufacturer.
- Select medium crust and white bread settings. You can use less yeast during the hot and humid months on summer.

NUTRITIONS

- Calories: 34 Cal
- Total Carbohydrate: 5.7 g
- Cholesterol: 1 mg
- Total Fat: 1 g
- Protein: 1 g
- Sodium: 313 mg

Chapter 12.

Gluten-Free Bread

Gluten-Free Chia Bread

This bread is excellent to eat with butter slathered all over it. It is great as breakfast, in the morning, with lunch or dinner.

Preparation:

5 minutes

Cooking Time:

3 Hour

Servings:

12

INGREDIENTS

- 1 cup warm water
- 3 large organic eggs, room temperature
- 1/4 cup olive oil
- 1 Tablespoon apple cider vinegar
- 1 cup gluten-free chia seeds, ground to flour
- 1 cup almond meal flour
- 1/2 cup potato starch
- 1/4 cup coconut flour
- 3/4 cup millet flour
- 1 Tablespoon xanthan gum
- 1 1/2 teaspoons salt
- 2 Tablespoons sugar
- 3 Tablespoons nonfat dry milk
- 6 teaspoons instant yeast

DIRECTIONS

- Mix wet mixtures and add to the bread maker pan. Whisk dry ingredients, except yeast, together and add on top of wet ingredients.
- Make a well in the dry ingredients, add yeast, select Whole Wheat cycle, light crust color, and press Start. Allow cooling completely before serving.

NUTRITIONS

- Calories: 375 Cal
- Fat: 18.3 g
- Carbs: 42 g
- Protein: 12.2 g

Easy Gluten-Free, Dairy-Free Bread

This bread is a beautiful, delicious, and easy gluten-free, dairy-free bread. It can be eaten for breakfast, dinner, or dessert.

Preparation:

15 minutes

Cooking Time:

2 Hours 10 Minutes

Servings:

12

INGREDIENTS

- 1 1/2 cups warm water
- 2 teaspoons active dry yeast
- 2 teaspoons sugar
- 2 eggs, room temperature
- 1 egg white, room temperature
- 1 1/2 Tablespoons apple cider vinegar
- 4 1/2 Tablespoons olive oil
- 3 1/3 cups multi-purpose gluten-free flour

DIRECTIONS

- Put the yeast plus sugar into the warm water and stir to mix in a large mixing bowl; set aside until foamy, about 8 to 10 minutes.
- Whisk the 2 eggs and 1 egg white together in a separate mixing bowl and add to the bread maker's baking pan. Put apple cider vinegar plus oil in the baking pan.
- Put foamy yeast/water batter into the baking pan. Add the multi-purpose gluten-free flour on top.
- Set for Gluten-Free bread setting and Start. Remove and move onto a cooling rack, then let cool completely before slicing to serve.

NUTRITIONS

- Calories: 241 Cal
- Fat: 6.8 g
- Carbs: 41 g
- Protein: 4.5 g

Gluten-Free Crusty Boule Bread

This bread is packed with flavor, crusty on the outside and soft on the inside. Great for sandwiches and toast. No more frozen bread; instead of fresh from the bread machine, this bread is ready to slice and go.

Preparation:

15 minutes

Cooking Time:

3 Hours

Servings:

12

INGREDIENTS

- 3 1/4 cups gluten-free flour mix
- 1 Tablespoon active dry yeast
- 1 1/2 teaspoons kosher salt
- 1 Tablespoon guar gum
- 1 1/3 cups warm water
- 2 large eggs, room temperature
- 2 Tablespoons, plus 2 teaspoons olive oil
- 1 Tablespoon honey

DIRECTIONS

- Mix all of the dry mixtures, except the yeast, in a large mixing bowl; set aside. Mix the water, eggs, oil, plus honey in a separate mixing bowl.
- Pour the wet ingredients into the bread maker. Put the dry mixtures on top of the wet ingredients.
- Make a well in the center of the dry fixing and add the yeast. Set to Gluten-Free setting and press Start.
- Remove baked bread and let cool completely. Hollow out and fill with soup or dip to use as a boule or slice for serving.

NUTRITIONS

- Calories: 480 Cal
- Fat: 3.2 g
- Carbs: 103.9 g
- Protein: 2.4 g

Gluten-Free Sorghum Bread Recipe

This bread is surprisingly good for gluten-free bread; it is very light and airy. Its crumb is very soft. It probably won't stand up to sandwich bread, but gluten-free people will love this bread. It makes a great sandwich loaf too.

Preparation:

5 minutes

Cooking Time:

3 Hours

Servings:

12

INGREDIENTS

- 1 1/2 cups sorghum flour
- 1 cup tapioca starch
- 1/2 cup brown sweet rice flour
- 1 teaspoon xanthan gum
- 1 teaspoon guar gum
- 1/2 teaspoon salt
- 3 Tablespoons sugar
- 2 1/4 teaspoons instant yeast
- 3 eggs (room temperature, lightly beaten)
- 1/4 cup oil
- 1 1/2 teaspoons vinegar
- 3/4-1 cup milk (105 - 115°F)

DIRECTIONS

- Mix the dry mixtures in a mixing bowl, except for the yeast. Put the wet mixtures in the bread maker pan, then add the dry ingredients on top.
- Create a well or hole in the middle of the dry mixtures and add the yeast. Set to Basic bread cycle, light crust color, and press Start. Remove and lay on its side to cool on a wire rack before serving.

NUTRITIONS

- Calories: 169 Cal
- Fat: 6.3 g
- Carbs: 25.8 g

- Protein: 3.3 g.

Gluten-Free Oat & Honey Bread

This bread is easy to make and keeps well for up to a week when stored in an air-tight container.

Preparation:

5 minutes

Cooking Time:

3 Hours

Servings:

12

INGREDIENTS

- 1 1/4 cups warm water
- 3 Tablespoons honey
- 2 eggs
- 3 Tablespoons butter, melted
- 1 1/4 cups gluten-free oats
- 1 1/4 cups brown rice flour
- 1/2 cup potato starch
- 2 teaspoons xanthan gum
- 1 1/2 teaspoons sugar
- 3/4 teaspoon salt
- 1 1/2 Tablespoons active dry yeast

DIRECTIONS

- Add ingredients in the order listed above, except for the yeast. Make a well in the center of the dry mixtures and add the yeast.
- Select a Gluten-Free cycle, light crust color, and press Start.
- Remove bread and allow the bread to cool on its side on a cooling rack for 20 minutes before slicing to serve.

NUTRITIONS

- Calories: 151
- Fat: 4.5 g
- Carbs: 27.2 g
- Protein: 3.5 g

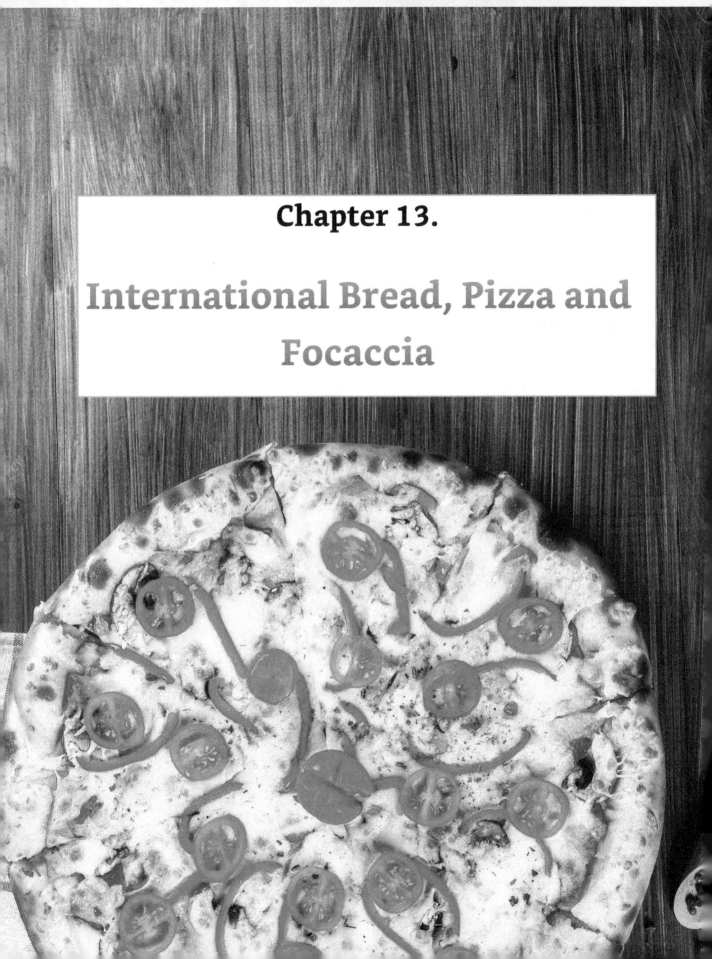

Chapter 13.

International Bread, Pizza and Focaccia

European Black Bread

This bread is well known to people in Northern Europe; particularly in the Nordic countries. The German name for the bread is "Schwarzbrot" (Black Bread), and it is very much a national bread of Sweden. It makes a delicious base for sandwiches and can make a variety of sweet and savory treats.

Preparation:

2 Hours

Cooking Time:

1 Hour 5 Minutes

Servings:

1 Loaf

INGREDIENTS

- ¾ teaspoon cider vinegar
- 1 cup water
- ½ cup rye flour
- 1 ½ cups flour
- 1 Tablespoon margarine
- ¼ cup oat bran
- 1 teaspoon salt
- 1 ½ Tablespoons sugar
- 1 teaspoon dried onion flakes
- 1 teaspoon caraway seed
- 1 teaspoon yeast
- 2 Tablespoons unsweetened cocoa

DIRECTIONS

- Put everything in your bread machine.
- Now select the basic setting. Hit the start button.
- Transfer bread to a rack for cooling once done.

NUTRITIONS

- Calories 114 Cal
- Carbohydrates: 22 g
- Fat 1.7 g
- Protein 3 g

Italian Bread

This bread is at once crisp and chewy with a beautiful golden-brown crust.

Preparation:

2 Hours

Cooking Time:

1 Hours 10 Minutes

Servings:

2 Loaves

INGREDIENTS

- 1 Tablespoon light brown sugar
- 4 cups all-purpose flour, unbleached
- 1 ½ teaspoons salt
- 1 1/3 cups + 1 Tablespoon warm water
- 1 package active dry yeast
- 1 ½ teaspoons olive oil
- 1 egg
- 2 Tablespoons cornmeal

DIRECTIONS

- Place flour, brown sugar, 1/3 cup warm water, salt, olive oil, and yeast in your bread machine. Select the dough cycle. Hit the start button. Deflate your dough. Turn it on a floured surface. Form two loaves from the dough.
- Keep them on your cutting board. The seam side should be down. Sprinkle some cornmeal on your board. Place a damp cloth on your loaves to cover them. Wait for 40 minutes. The volume should double.
- In the meantime, preheat your oven to 190°C. Beat a 1 Tablespoon of water and an egg in a bowl. Brush this mixture on your loaves.
- Make an extended cut at the center of your loaves with a knife. Shake your cutting board gently, making sure that the loaves do not stick. Now slide your loaves on a baking sheet— Bake in your oven for about 35

minutes.

NUTRITIONS

- Calories: 105 Cal
- Carbohydrates: 20.6 g
- Fat: 0.9 g
- Protein: 3.1 g

Pita Bread

Pita bread is a soft pillowy bun with a crisp outside and soft inside. It can be eaten with any type of meal.

Preparation:

35 Minutes

Cooking Time:

20 Minutes

Servings:

8

INGREDIENTS	DIRECTIONS
<ul><li>3 cups all-purpose flour</li><li>1 1/8 cups warm water</li><li>1 Tablespoon vegetable oil</li><li>1 teaspoon salt</li><li>1 ½ teaspoons active dry yeast</li><li>1 active teaspoon white sugar</li></ul>	<ul><li>Place all the mixtures in your bread pan. Select the dough setting. Hit the start button. The machine beeps after the dough rises adequately.</li><li>Turn the dough on a floured surface. Roll, then stretch the dough gently into a 12-inch rope. Cut into eight pieces with a knife. Now roll each piece into a ball. It should be smooth.</li><li>Roll each ball into a 7-inch circle. Keep covered with a towel on a floured top for 30 minutes for the pita to rise. It should get puffy slightly.</li><li>Preheat your oven to 260°C. Keep the pitas on your wire cake rack. Transfer to the oven rack directly.</li><li>Bake the pitas for 5 minutes. They should be puffed. The top should start to brown. Take out from the oven. Keep the pitas immediately in a sealed paper bag. You can also cover using a damp kitchen towel.</li><li>Split the top edge or cut it into half once the pitas are soft. You can also have the whole pitas if you want.</li></ul>

NUTRITIONS

- Calories 191 Cal
- Carbohydrates: 37g
- Fat 3g
- Protein 5g

Sour Cream Chive Bread

This bread is excellent for campfire additions like chevre, swiss, or gorgonzola. Experiment with different cheese types and seasonings for other flavors, like chilies, for instance.

Preparation:

10 Minutes

Cooking Time:

3 hours

Servings:

1 Loaf

INGREDIENTS

- 2/3 cup whole milk at 70°-80°F
- 1/4 cup water at 70°-80°F
- 1/4 cup sour cream
- 2 Tablespoons butter
- 1-1/2 teaspoons sugar
- 1-1/2 teaspoons salt
- 3 cups bread flour
- 1/8 teaspoon baking soda
- 1/4 cup minced chives
- 2-1/4 teaspoons active dry yeast

DIRECTIONS

- Place all the mixtures in the bread machine pan.
- Select the white bread or basic setting and the medium crust setting.
- Hit the start button. Take out your hot loaf once it is done. Keep on your wire rack for cooling.
- Slice your bread once it is cold and serve.

NUTRITIONS

- Calories 105 Cal
- Fat 2g
- Carbohydrate 18g
- Protein 4g

Ethiopian Milk and Honey Bread

Ethiopian Milk and Honey Bread is a tasty treat for breakfast, after a meal when you are having coffee or tea.

Preparation:

120 Minutes

Cooking Time:

1 Hour 15 Minutes

Servings:

1 Loaf

INGREDIENTS

- 3 Tablespoons honey
- 1 cup + 1 Tablespoon milk
- 3 cups bread flour
- 3 Tablespoons melted butter
- 2 teaspoons active dry yeast
- 1 ½ teaspoons salt

DIRECTIONS

- Add everything to the pan of your bread. Select the white bread or basic setting and the medium crust setting.
- Hit the start button. Take out your hot loaf once it is done. Keep on your wire rack for cooling.
- Slice your bread once it is cold and serve.

NUTRITIONS

- Calories 129 Cal
- Carbohydrates: 20 g
- Fat 3.8 g
- Protein 2.4 g

Chapter 14.

NEW RECIPES

1.Kolaches

Preparation:

6 minutes

Cooking Time:

15 minutes

Servings:

8 : Yield

INGREDIENTS

- 3 7/8 cups bread flour
- ¼ cup instant mashed potato flakes
- 1/3 cup milk powder
- ½ cup butter melted
- ½ cup butter softened to room temperature
- 1 egg yolk
- 1 egg
- ¼ cup white sugar
- 2 tsp dry active yeast
- 1 ¼ cups warm water
- 1 tsp salt
- 1 can 12 oz poppy seed filling
- 1 can 12 oz cherry pie filling

DIRECTIONS

- 1.Place the flour, mashed potato flakes, softened butter, sugar, yeast, egg yolk, milk powder and a dash of salt in the pan of a bread machine.
- 2.Crack the egg into it, followed splashes of warm water and set the dough cycle before running the machine to mix the ingredients together to form a dough; knead in another tbsp or two of water after 5 minutes of processing the ingredients if required.
- 3.Scoop out 1 tablespoon-worth of dollops of dough onto a cookie sheet, greased previously with a knob of softened butter, and cover them loosely with a plastic sheet.
- 4.Set aside in a warm spot for an hour to allow them rise to twice their previous size.

- 5.Flatten them out between the palms of your hands and make a small well in the centre of each dough ball, using your thumb.
- 6.Spoon a tbsp of the filling into the well of each dough ball and set them aside, covered, for another half an hour to raise again.
- 7.Meanwhile, warm up the oven to 375 degrees F.
- 8.Bake the pastries for 13-15 minutes or until they are just browned.
- 9.Smother them with a lashing of melted butter to glaze and set aside until they cool down to room temperature.

NUTRITIONS

- Calories: 771 Cal
- Total Carbohydrate: 80 g
- Total Fat: 44 g
- Protein: 18 g

2. Good Morning Sun Lemon Blueberry

Preparation:

25 minutes

Cooking Time:

62 minutes

Servings:

12 Slices

INGREDIENTS

- 1 c of granulated sugar
- 2 c all purpose flour
- 1/2 t salt, I used table salt
- 2 tsp baking powder
- 2 eggs
- 1/2 c milk
- 1/2 c butter, melted (Kerrygold is my favorite)
- zest of one medium lemon
- 1 1/2 c fresh blueberries, washed

DIRECTIONS

- 1.Spray the inside of a loaf pan with cooking spray. I like this coconut oil spray.
- 2.Preheat your oven to 350 degrees.
- 3.In a large bowl add all of the dry ingredients; granulated sugar, flour, salt and baking powder.
- 4.Use a zester and grate into the dry ingredients bowl the zest of the entire lemon. If you don't want a lot of lemon flavor you can just do half or less.
- 5.If you don't have a lemon you can certainly do lemon extract instead.
- 6.Use a spoon to mix those dry ingredients and zest together.
- 7.Set the bowl aside.
- 8.In a small bowl, crack and whisk your eggs.
- 9.Add in the milk and whisk it

into the eggs.

- 10. Pour the egg and milk mixture into the flour mixture. Begin to combine it.
- 11. Now you can add the melted butter. I don't like to add melted butter into the eggs because you do not want it to cook the eggs, ideally. If you want to add the butter to the egg and milk mixture you can, but just be mindful especially if it is really hot butter.
- 12. Combine the batter. I like to just use a large wooden spoon.
- 13. Now that all of this is combined it will be easy to gently fold the fresh blueberries in. Be gentle you don't want to smash the berries.
- 14. Pour the batter into the prepared bread pan and pop it in the oven for sixty minutes or until a toothpick inserted in the middle comes out clean.

NUTRITIONS

- Calories: 239 Cal
- Total Carbohydrate: 37 g
- Total Fat: 9 g
- Protein: 4 g

3.Polynesian Bread

INGREDIENTS

- 1 cup milk
- 1 egg
- 2 tablespoons butter
- 3/4 teaspoon salt
- 3 drops lemon essential oil
- 3 cups bread flour
- 1/3 cup white sugar
- 2.5 teaspoons active dry yeast

DIRECTIONS

- .Combine the flour and sugar in a bowl.
- .Add the milk, butter, egg, lemon essential oil and salt to the bottom of your bread machine pan.
- .Dump the flour mixture on top of the wet ingredients in the pan.
- .Pour the yeast on the top of the flour.
- Set the bread machine settings to 2 lb loaf, light crust, and Sweet Bread and START.

NUTRITIONS

4. Easy Fluffy Bread

Preparation:

25 minutes

Cooking Time:

25minutes

Servings:

1 Loaf

INGREDIENTS

- 1 cup hot water
- 2 teaspoons yeast
- 3 tablespoons sugar
- /cup vegetable oil
- 1 teaspoon salt
- 3 cups white flour

DIRECTIONS

- 1.Put your hot water, yeast and sugar in your bread machine. Let stand 6-12 minutes
- 2.depending on the temp of your water. Your yeast will foam.
- 3.Add remaining ingredients.
- 4.Use rapid or basic white cycles.
- 5.Or bake for 25 minute 350.

NUTRITIONS

- Calories: 2019.1 Cal
- Total Carbohydrate: 327.2 g
- Total Fat: 58.8 g
- Protein: 42 g

5. Tuscan Green Bread

Preparation:

6 minutes

Cooking Time:

215 minutes

Servings:

1 Loaf

INGREDIENTS

- 3 Tbsp Olive Oil
- 1 tsp Dried Rosemary
- 1/8 tsp Dried Ground Thyme
- 1/2 tsp Dried Basil Leaves
- 1/2 tsp Dried Oregano Leaves
- 1 cup Luke Warm Water
- 1 1/2 tsp Sea Salt
- 1/2 tsp Garlic Powder
- 3 Tbsp White Sugar
- 2 1/2 tsp Instant Dry Yeast
- 3 cups All-Purpose Flour

DIRECTIONS

- 1. Get your bread machine out and plug in. In a small bowl stir your dried herbs and olive oil together and set aside.
- 2. Then pour the ingredients into the basket of the bread machine preferably in this order; Water, Salt, Sugar, Yeast, Olive Oil mixture and then Flour.
- 3. Close the lid; set your machine to French setting, 1.5 lb loaf and light crust. Click start and you'll have beautiful fresh flavorful bread in about 3 1/2 hours.
- 4. When the bread is done, carefully remove it from the basket and let sit on counter for 5 min.
- 5. Best enjoyed warm; plain, with butter, olive oil or a dip.

6.Easy White bread

Preparation:

55 minutes

Cooking Time:

25 minutes

Servings:

1 Loaf

INGREDIENTS

- 1 cup milk, warmed
- 1 1/2 eggs, at room temperature
- 2 1/2 tablespoons butter, softened
- 1/4 cup sugar
- 3/4 teaspoon salt
- 3 cups bread flour
- 2 1/2 teaspoons yeast

DIRECTIONS

- 1.Combine all the ingredients in the bread machine according to the manufacturer's directions.
- 2.Use the basic or white bread setting. Adjust the crust setting to "light".
- 3.When baked, let the bread cool in the pan for 2-3 minutes then remove from the pan to a wire rack to cool completely before slicing. Store in an airtight container.

NUTRITIONS

- Calories: 123 Cal
- Total Carbohydrate: 20 g
- Total Fat: 3 g
- Protein: 4 g

7. Perfect Sticky Buns Old School

Preparation:

3 hour 5 minutes

Cooking Time:

25 minutes

Servings:

6 Servings

INGREDIENTS

- DOUGH
- 1/2 tablespoon yeast
- 3 cups bread flour
- 1/4 cup sugar
- 3/4 teaspoon salt
- 2 tablespoons oil
- 1 egg, room temp
- 1/2 cup warm milk
- 1/2 cup warm water
- TOPPING
- 1/2 cup brown sugar
- 1/2 cup nuts, chopped
- 1/4 cup melted butter
- 2 tablespoons corn syrup
- FILLING
- 1/4 cup butter, melted

DIRECTIONS

- 1.Place dough ingredients into bread machine pan in order listed. Select manual or dough setting.
- 2.Mix sticky topping ingredients and spread evenly on bottom of nonstick 13"x9" or 9" round pan. After the machine has completed the second rising, it will beep.
- 3.On a lightly floured surface, roll or pat to 12"x9" rectangle. Brush with melted butter. Combine sugars, nuts, and cinnamon; sprinkle over dough. Starting with shorter side, roll up tightly. Pinch edge to seal. Cut into 12 slices.
- 4.Place on greased 13"x9" pan. Or cut 9 slices, throw away the two end pieces and place in 9" round pan, with one in the middle and the rest arranged

- 3 tablespoons sugar
- 3 tablespoons brown sugar
- 1/4 cup nuts, chopped
- 1 tablespoon ground cinnamon

around the outside. (looks real nice) Cover; let rise in a warm place until almost double in volume, (about 30 minutes).

- 5.Preheat oven to 375 degrees F. Bake 20-26 minutes.
- 6.Carefully turn out rolls and serve immediately.

NUTRITIONS

- Calories: 701 Cal
- Total Carbohydrate: 92 g
- Total Fat: 33 g
- Protein: 13 g

8.Odin Rye Bread

Preparation:

60 minutes

Cooking Time:

30 minutes

Servings:

16 Servings

INGREDIENTS

- 1 cup water
- 1/3 cup molasses
- 2 1/2 tablespoons butter or margarine
- 1/2 teaspoon salt
- 1 tablespoon caraway seed, optional
- 1/3 cup whole wheat flour
- 1 2/3 cup rye flour
- 1 2/3 cup bread flour
- 2 teaspoons yeast

DIRECTIONS

- 1.Put ingredients in bread machine in the order suggested by the manufacturer and bake on regular cycle.
- 2.There is a large amount of flour in this recipe. If the dough seems dry (check about 5 minutes into the cycle) or the machine sounds like it is struggling, add water 1 tablespoon at a time.

NUTRITIONS

- Calories: 154 Cal
- Total Carbohydrate: 29 g
- Total Fat: 3 g
- Protein: 4 g

9.Polish Bagels

Preparation:

60 minutes

Cooking Time:

25 minutes

Servings:

9 Servings

INGREDIENTS

- 1 cup warm water (110 degrees F)
- 1 1/2 teaspoon salt
- 2 tablespoons white sugar
- 3 cups bread flour
- 2 1/4 teaspoons active dry yeast
- 3 quarts boiling water
- 3 tablespoons white sugar
- corn meal
- egg white
- poppy seeds
- dried onions

DIRECTIONS

- Place water, salt, sugar, flour, and yeast in the bread machine pan according to the order recommended by the manufacturer. Select dough setting.
- When dough cycle is complete, let dough rest on a lightly floured surface. Meanwhile, in a large pot bring 3 quarts of water to a boil. Stir in 3 tablespoons of sugar.
- Cut dough into 9 equal pieces and roll each piece into a small ball. Flatten balls. Poke a hole in the middle of each with your thumb. Twirl the dough on your finger or thumb to enlarge the hole and even out the dough around the hole. Cover bagels with a clean cloth and let rest for 10 minutes.
- Carefully transfer bagels to

boiling water. Boil for 1 minute, turning half way through. Drain briefly on clean towel. Move boiled bagels to baking sheet sprinkled with corn meal.

- Glaze tops with egg white and sprinkle with your choice of toppings. Bake in a preheated 375 degrees F oven for 20 to 25 minutes, until well browned.

NUTRITIONS

- Calories: 195 Cal
- Total Carbohydrate: 40 g
- Total Fat: 1 g
- Protein: 6 g

10.Easter Raisins Bread

Preparation:

4 hour 5 minutes

Cooking Time:

60 minutes

Servings:

1 Loaf

INGREDIENTS

- 1 cup whole milk, warmed
- 2 eggs, lightly beaten
- 3 tablespoons butter, melted
- 1/2 cup granulated sugar
- 1 teaspoon salt
- 3 cups bread flour
- 1 packet (2.25 teaspoon) dry active yeast
- 2/3 cup raisins, optional

DIRECTIONS

- 1.Combine the ingredients in a bread machine as directed by the machine manufacturer. Do not use a delay start setting. Set the machine to basic white bread with a light crust.
- 2.When the bread has finished the baking cycle, remove the pan from the machine and let the bread cool in the pan for 5 minutes.
- 3.Remove the bread from the pan and let cool completely on a wire rack before slicing or storing in an airtight container.

NUTRITIONS

- Calories: 208 Cal
- Total Carbohydrate: 34 g
- Total Fat: 5 g
- Protein: 6 g

11.Easy French Bread

Preparation:

2 hour 55 minutes

Cooking Time:

35 minutes

Servings:

16 Yield

INGREDIENTS

- 1 cup lukewarm water (227 gr)
- 1-1/2 teaspoons sugar (8 gr)
- 1-1/2 teaspoons salt (9 gr)
- 1-1/2 teaspoons butter (8 gr)
- 3 cups bread flour (360 gr)
- 1 teaspoon instant yeast (3 gr)
- Glaze:
- 1 egg white (35 gr)
- 1 teaspoon water (5 gr)

DIRECTIONS

- 1.Place all ingredients in a bread machine pan in the order listed. Select the DOUGH cycle and press START. After 5-10 minutes, lift the lid and check thedough. If the dough is too sticky (levels out), add more flour 1 tablespoon at a time. If the dough is too dry, add water a tablespoon at a time. The doughshould be a loose ball that sticks to the sides, then pulls away cleanly.
- 2.When the dough cycle is complete, turn the dough out onto a lightly floured board. Roll into an oval shape slightly longer than you want your final loaf to be (about 9 x 12 inches is what I do.)
-
- 3.Starting from a long side, roll into a cylinder shape. Pinch seam together. Pinch ends together. Pull ends to reach seams and pinch together with the seam making a small rounded shape

on each end. (This is difficult to describe. See the picture above or watch the video if you are confused.)

-

- 4.Flip the loaf over and carefully place it onto a greased baking sheet or one covered with a silicone baking sheet or parchment paper. Cover with a tea towel and allow to rise in a warm place until almost double.

- 5.Preheat oven to 425 °F.

- 6.Whisk egg white and water together for glaze. Brush over loaf.

- 7.Using a sharp knife, (I use a serrated knife) or a single-edge razor blade, cut diagonal slashes about 2 inches apart and 1/2 inch deep across the top of the loaf.

- 8.Bake loaf for 20 minutes. Reduce heat to 350 degrees and bake another 5- 10 minutes until golden brown. Bread should reach 190 °F in the middle when cooked through.

- 9.Remove to a cooling rack or slice and eat immediately.

NUTRITIONS

- Calories: 43 Cal
- Total Carbohydrate: 7 g
- Total Fat: 1 g
- Protein: 1 g

12. Sunflower and Oatmeal Bread

Preparation:

3 hour 28 minutes

Cooking Time:

30 minutes

Servings:

1 Loaf

INGREDIENTS

- 1/2 cup milk
- 1/2 cup water
- 1/4 cup honey
- 2 tablespoons unsalted butter
- 1 1/4 teaspoon salt
- 3 cups bread flour
- 1/2 cup (50 grams) quick or old-fashioned oats (not instant) (45 gr)
- 2 and 1/4 teaspoons bread machine or instant yeast
- 1/2 cup hulled sunflower seeds, toasted

DIRECTIONS

- 1.Warm milk and water in the microwave for one minute on HIGH.
- 2.Add to bread machine pan along with remaining ingredients except seeds in order given.
- 3.Select "Dough" cycle and start. After about 5-10 minutes, lift the lid and add extra liquid or extra flour 1 tablespoon at a time, if necessary, to correct consistency. Dough should stick to side of pan, then pull away.
- 4.Add the seeds at the Raisin/Nut signal or 5-10 minutes before the kneading cycle ends. If you miss it, you can always work them in by hand when you get ready to form the loaf.
- 5.When dough cycle has completed, remove dough to a

floured surface and flatten into a rectangle. Roll into a cylinder. Place into a 9x5-inch greased loaf pan with the seam down and tucking the ends under.

- 6.Loosely cover (I use a shower cap or tea towel) and set in a warm place until dough rises approximately 1/2 to 1 inch above the rim of the pan.
- Preheat oven to 350 degrees.
- 1.Bake for 30-35 minutes or until internal temperature reaches 190 degrees F. I suggest you test it with a thermometer if you are a novice bread baker.
- 2.Note: Check loaf half way through baking and cover with foil if getting too brown.

NUTRITIONS

- Calories: 210 Cal
- Total Carbohydrate: 34 g
- Total Fat: 6 g
- Protein: 6 g

13. Easy Milk Bread

Preparation:

2 hour 35 minutes

Cooking Time:

30 minutes

Servings:

12 Servings

INGREDIENTS

- 8 ounces water (room temperature)

- 1/2 cup sweetened condensed milk

- 1 teaspoon salt

- 1 tablespoon butter (room temperature)

- 3 cups (+) bread flour

- 2 scant teaspoons instant yeast

DIRECTIONS

- 1.Add ingredients to the bread pan in the order listed.
- 2.Select the Dough Cycle and start.
- 3.Raise the lid and check dough after about 10 minutes. Add flour one tablespoon at a time, if necessary, until dough reaches the correct consistency. It should come together in a ball that sticks to the side of the pan, then pulls away cleanly. If dough thumps against the side of the pan, add warm water 1 tablespoon at a time. If the dough is sticky and doesn't pull away from the side, add flour 1 tablespoon at a time until the dough starts to form a slightly sticky ball.
- 4.Remove dough from the pan at the end of the dough cycle and place onto a lightly floured board. Knead by hand a little bit to press out any large air bubbles.

- 5.Roll into a rectangle approximately 9 x 11 inches. Roll up starting from the long end, and tuck ends to fit into greased 9 x 4-inch loaf pan. Let rise until the dough is doubled from its original size. Because this dough is a "high-riser," be careful not to let the dough rise too much or it will cave in on the sides and/or the top.
- 6.Preheat oven 15 minutes before you estimate your loaf will be ready.
- 7.Bake at 375°F for 35-45 minutes. The Interior should reach 190 degrees. Place a foil tent over bread halfway through baking to protect from over-browning.
- 8.Allow cooling for 15 minutes before turning out to cool completely. It's best if you wait at least two hours before slicing so the loaf will hold its shape without squishing under the pressure of a knife.

NUTRITIONS

- Calories: 175 Cal
- Total Carbohydrate: 32 g
- Total Fat: 3 g
- Protein: 5 g

14.Big Crusty Bread

Preparation:

18 hour 32 minutes

Cooking Time:

30 minutes

Servings:

12 Servings

INGREDIENTS	DIRECTIONS

INGREDIENTS

- Sponge:
- 1-1/2 cups (180 grams) all-purpose, unbleached flour
- 1 teaspoon instant or bread machine yeast
- 1 cup (8 oz) water

Dough:

- 3 tablespoons (1.5 oz) water
- 1 teaspoon sugar (optional)
- 1-1/2 teaspoon salt
- 1-3/4 cups (210 grams) all-purpose, unbleached flour

DIRECTIONS

Making the Sponge:

1.Place water, yeast and flour in bread machine pan and select the "dough" cycle. Allow to mix about 5 minutes using small spatula to carefully push flour stuck in the corners into the mixing area. Unplug machine and let stand at room temperature over night or about 8 hours. Do not leave over 16 hours.

Making the Dough

2.Open lid of bread machine and add water, sugar, salt, and flour.

3.Restart dough cycle. Check dough after 5-10 minutes of mixing. If necessary, add additional flour 1 tablespoon at a time to form a smooth but slightly tacky ball or water if dough is too dry and bounces against the sides.

4.When dough cycle ends, allow dough to continue to rise in machine for at least 30 minutes (or more if ambient temperature is cool) until double in size. If you are new to bread machines, see Six Bread Machine Tips for Beginners for more help with this step.

Preparing and Baking the Loaf

5.Remove dough from bread machine pan to lightly floured board or silicone baking mat (my preference). Form into smooth ball by pulling dough around to bottom until top is smooth. Place on parchment-covered cookie sheet. Cover loosely with lightly oiled plastic wrap and place in warm place to rise until almost double.

6.About 15 minutes before bread is ready to bake, preheat oven to 425 degrees. Just before putting bread in the oven, sprinkle top with flour. Using a single edge razor blade (or a sharp, serrated knife), make several cuts across top of bread about 1/2 inch deep.

7.Bake 30-35 minutes until loaf is golden brown and internal temperature has reached 190 degrees. Allow to cool on rack before slicing. Or slice while it's hot at the risk of squashing your bread. It's worth it.

NUTRITIONS

- Calories: 78 Cal
- Total Carbohydrate: 16 g
- Total Fat: 0 g
- Protein: 2 g

15.Sweet Brioche

Preparation:

125 minutes

Cooking Time:

16 minutes

Servings:

12 Rolls

INGREDIENTS

- 1/3 cup milk (82 g)

- 1 tablespoon bread flour (9 g)

- 3 large eggs, room temperature (150 gr)

- 1-1/4 teaspoon salt (7.5g)

- 2 tablespoons sugar (25 g)

- 2-3/4 cups bread flour (330 g)

- 2-1/4 teaspoons bread-machine or instant yeast (7 gr)

- 12 tablespoons butter, pliable (IMPORTANT) (168 gr)

- Glaze:

- 1 egg (50 gr)+ 1 tablespoon heavy cream whisked together well.

DIRECTIONS

- 1.Combine 1/3 cup of milk and 1 tablespoon of flour in a microwave-safe container. Whisk until smooth.

- 2.Microwave this liquid paste mixture for 20 seconds on High. Remove and stir. Place back into the microwave for 10-20 seconds or however long it takes to turn the mixture into a thick "gravy" consistency. Pour into the bread machine pan.

- 3.Add eggs, sugar, salt, 2-3/4 cups of bread flour, and yeast.

- 4.Select the DOUGH cycle and push start. After 8 minutes, open the lid of your bread machine and observe the kneading action. If the dough is too slack, add additional flour one tablespoon at a time, letting the dough absorb the flour before adding more.

- 5.You want the dough to be thick enough to hold its shape, stick to the sides, then pull away, but not "cleanly." This dough should be stickier than the

(14 gr)

average bread dough. But it must be firm enough for the bread machine blades to get traction as they knead the dough.

- 6.About 15 minutes into the dough cycle, open the lid again. Begin to add the butter to the dough, one tablespoon at a time. Let the dough absorb each piece of butter before adding more. The dough should be smooth and shiny at this point and pulling away from the sides.
- 7.Allow dough cycle to complete. The dough should be doubled in size. If the ambient temperature is chilly in your kitchen, you may need to allow the dough to rise longer until doubled.
- 8.Gently release the dough from the sides to remove some of the air.
- 9.Cover the bread machine bowl and place into your refrigerator for 6-24 hours. Do not skip this part. If you don't have time for the chill, you might want to make another kind of bread.
- Instructions for shaping burger buns:
- 1.Form dough into 2 logs.
- 2.Cut each log into 5 equally-sized pieces. If you want sliders, cut more pieces. You get to decide. Make more than one size if you want to please everybody.
- 3.Form each portion into a ball and flatten it somewhat with your fingers

on the Silpat or parchment-paper-lined baking tray.

- 4.Cover each bun with plastic wrap and smash it. I like to use a transparent glass pie plate so I can see how evenly I'm smashing the bun. Cover with a tea towel and allow to proof.
- 5.Preheat oven to 425 degrees F.
- 6.After rising to almost double their original size, press each bun gently and evenly with your fingers. Don't worry. They pop right up once you put them in the oven.
- 7.Paint with the glaze. Turn oven temperature back to 350 degrees F and bake for 15 minutes.
- 8.Remove buns onto a cooling rack. Slice horizontally with a serrated knife to use as buns.
- Shaping a classic brioche with a topknot:
- 1.Place dough on a lightly floured board. LIghtly knead and mold into a ball. Divide in half. Cut each half into 6 pieces.
- 2.Pull a small amount off each of the 12 balls to make hats. Roll all portions into little balls. The smoother the better and practice helps. Place one large ball in each mold or fill a muffin tin. Place all small balls (future hats) on wax paper, parchment, or a silicone mat on cookie sheet.

- 3.Cover rolls with a tea towel and allow them to rise in a warm place until almost doubled. This may take 1-2 hours.
- 4.When rolls have almost doubled in size, use a greased thumb or the handle end of a wooden spoon to carefully depress dough in the center (all the way to the bottom.) Don't worry, it will spring back once it hits the oven. Brush with glaze.
- 5.Place a small ball in the center of the roll and again brush entire roll with glaze, taking care not to let glaze pool at the edges between the dough and the mold.
- 6.Place individual molds or muffins pans onto a cookie sheet to keep the bottoms from over-browning.
- 7.Preheat oven to 425 degrees. Then reduce temperature to 375 degrees and bake rolls for about 15 minutes. Loosely cover rolls with foil if tops are getting too dark. Internal temperature should reach 185-190 degrees.
- 8.Allow the rolls to cool for a couple of minutes. Turn out onto a cooling rack.
- 9.Best eaten the same day but also good toasted the next day.
- Shaping a brioche loaf:
- 1.Roll out dough into an 11 x 15-inch rectangle with the short side towards you.

- 2.Roll up dough, starting with the short side. Divide into four equal parts.
- 3.Place each roll perpendicular to the long side of a greased 9 x 5 loaf pan. (I use Baker's Joy.)
- 4.Cover with a tea towel or shower cap. Allow dough to rise until it reaches the top of the pan.
- 5.Preheat oven to 425 degrees F.
- 6.Brush loaf with glaze.
- 7.Set oven temperature back to 350 degrees F. Bake for 30-35 minutes or until internal temperature reaches 190 degrees F. Cover loosely with a piece of aluminum foil if the top starts to get too brown.
- 8.Let the bread cool for about 10 minutes before removing it from the pan to a cooling rack.

NUTRITIONS

- Calories: 259 Cal
- Total Carbohydrate: 26 g
- Total Fat: 14 g
- Protein: 6 g

16.Fluffy Lemon and Cramberry Bread

Preparation:

175 minutes

Cooking Time:

15 minutes

Servings:

16 Rolls

INGREDIENTS

- 1 cup jellied cranberry sauce, room temperature
- 1/4 cup whole milk or half and half, lukewarm
- 1 egg
- 1 teaspoon salt
- 1/4 cup butter, softened
- Grated rind from two medium lemons
- 3 cups all-purpose unbleached flour
- 2 teaspoons bread machine yeast
- 1 cup dried cranberries

DIRECTIONS

- 1.Add all ingredients to the bread machine pan in the order given.
- 2.Select dough cycle and push the start button.
- 3.When the machine beeps to indicate the best time for additions, add dried cranberries to the dough.
- 4.When the dough cycle completes, remove dough from the bread machine and place it on a floured surface. (I prefer to use a silicone baking mat so I can throw it into the dishwasher when I'm done.)
- 5.Divide dough into two equal pieces. Then divide each of those dough balls into 8 equal-sized portions and shape them into balls.
- 6.Place balls into two 8-inch cake pans as pictured above.

- 7.Cover rolls with a tea towel or shower caps and let rolls rise for approximately 45 minutes or until almost double in size.
- 8.Preheat oven to 350˚F.
- 9.Bake rolls approximately 14-16 minutes or until golden brown.
- 10.After removing rolls from the oven, let them sit in the pan for about 5-8 minutes. Turn out onto a wire rack to cool. If you leave the hot rolls in the pan, they will become soggy on the bottom.
-

NUTRITIONS

- Calories: 182 Cal
- Total Carbohydrate: 35 g
- Total Fat: 2 g
- Protein: 3 g

Chapter 15.

Cooking Measurement Conversion

US Dry Volume Measurements	
1/16 teaspoon	a dash
1/8 teaspoon	a pinch
3 teaspoons	1 tablespoon
¼ cup	4 tablespoons
1/3 cup	5 tablespoons + 1 teaspoon
½ cup	8 tablespoons
¾ cup	12 tablespoons
1 cup	16 tablespoons
1 pound	16 ounces

US Liquid Volume Measurements	
Eight fluid ounces	1 cup
1 pint = 2 cups	16 fluid ounces
1 quart = 2 pints	4 cups
1 gallon = 4 quarts	16 cups

Grocery Shopping List

Grains and flours

Wheat Flour

- Semolina

- Corn Flour

- Corn Flakes

- Polenta

- Bread Flour

- Oatmeal Flour

- Oatmeal Flakes

- Wholemeal Flour

- Whole-grain Flour

- Grain Flour

- Rye Flour

- Rice Flour

- Flax Flour

- Bulgur

- Rice

- Oatmeal

- Bran

- Oat Bran

- Yeast

- Eggs

Nuts and Seeds

- Poppy Seeds

- Almond

- Almond Flakes

- Peanuts

- Sunflower Seeds

- Flaxseeds

- Mustard Seeds

- Sesame Seeds

- Pumpkin Seeds

- Nuts

- Walnut

- Dried Coconut

- Pistachios

- Pecans

- Hazelnut

- Sesame

Milk

- Milk

- Sour Cream

- Yogurt

- Fruit Yogurt

- Butter

- Whey/Serum

- Milk Powder

- Coconut Milk

Meat, Poultry & Fish

- Chicken

- Smoked Sausages

- Bacon

- Pepperoni

- Ham

- Smoked Fish

Cheese

- Parmesan

- Blue Cheese

- Gren Cheese

- Feta

- Cottage Cheese

- Mozzarella

- Cheddar

Canned Goods

- Olives

Dried Herbs, Vegetables & Spices

- Italian Herbs

- Oregano

- Basil

- Rosemary

- Dill

- Saffron

- Lavender

- Garlic Powder

- Thyme

- Dried Carrots

- Dried Beets

- Dried Onion

- Dried Garlic

- Sun-dried Tomatoes

Oils, Vinegar, & Condiments

- Extra-virgin Olive Oil

- Flaxseed Oil

- Rapeseed Oil

- Almond Oil

- Sunflower Oil

- Mustard Oil

- Corn Oil

- Peanut Butter

- Horseradish

- Soy Sauce

- Tomato Paste

- Tomato Sauce

- Tomato Sauce for Pizza

- Dijon Mustard

Beverages

- Rum

- Light Beer

- Dark Beer

- Instant Coffee

Seasonings

- Salt

- Sea Salt

- Sugar

- Brown Sugar

- Powdered Sugar

- Sugar Vanilla

- Vanillin

- Cane Sugar

- Baking Powder

- Pininfarina

- Cocoa Powder

- Molasses

- Ground Black Pepper

- Paprika

- Parsley

- Dill

- Spinach

- Garlic

- Lemon

- Cardamon

- Rosemary

- Nutmeg

- Cinnamon

- Dried Kvass

- Cumin

- Dry Mustard

- Granular Mustard

- Turmeric

Fruits & Vegetables

- Apple

- Mushrooms

- Cherry Tomatoes

- Bell Pepper

- Onion

- Carrots

- Beet

- Zucchini

- Cabbage

- Potato

- Bananas

- Orange

- Mandarin

- Pineapple

- Peach

- Apricot

- Strawberry

Sweets

- Honey

- Chocolate

- Dark Chocolate

- Dates

- Figs

- Raisings

- Prunes

- Cocktail Cherry

- Peanut Butter

- Marzipan

- Candied Fruits

- Dried Cherry

- Dried Apricot

- Dried Cranberry

- Dried Pineapple

- Berry Syrup

Flour, Yeast, Water, and Salt Work Together to Form the Bread

The combination of these simple elements is the base for our bread, but it is necessary to know these essential ingredients thoroughly to achieve a good result.

Let's start with the flour. It is an entire world, but we will say that it is nothing more than the result of the grinding of a cereal, seed, or tuber to simplify things a little. The most common flours for baking are wheat flour (there are many varieties of this cereal); oats; corn; rye; barley, and even nuts like chestnut. As the flour that is used more frequently is wheat, we will focus on it (although without stopping - for now - in its varieties).

In general terms, we could say that wheat flour is composed of starch and other elements (in variable proportions) such as minerals, vitamins, proteins, and ashes. The sifting of the milling influences these factors. The whole-grain, keeping the bran, make up the whole flours; if they are deprived of it, we will obtain white flours. There are flours of soft wheat and durum wheat, the difference of these lying in the amount of protein that each contains and, therefore, the result of the bread will be different.

The proteins (gliadin and glutenin) in the flour are the main things responsible for the dough's formation and elasticity that, together with the fermentation, make the bread have volume and consistency. As the flour is hydrated, the proteins bind, transforming into gluten. When we manipulate a bread dough, and they are oxygenated, the dough becomes elastic and workable. If the mass is well hydrated and kneaded, a protein mesh (glutinous network) is created that covers it. The more protein the flour has, the more water it will need, so you must be careful not to overdo it.

Yeast is the second great protagonist of bread. Typically, a fungus is used (suitable for ingestion), which can be found in two versions: dry (lyophilized) or fresh. Keep in mind that this latest version is a living organism, so it must be appropriately conserved, as it loses strength over time. If dry is used, the proportion of yeast is 1/3 of the fresh recipe's amount. For example, if it calls for 10 gr. of fresh yeast, you should use 3 gr. of dry yeast.

Another way to produce bread is natural sourdough, the oldest way to make the bread ferment (through bacteria present in the environment). When the bread is made with sourdough, it usually has a slightly acidic taste, lasts longer, has an intense smell, and facilitates digestion due to bacterial fermentation. Making the sourdough is simple, but it takes time (it usually takes about five days). Here you have a good recipe.

Water and salt have no significant complications and secrets. In fact, in the best professional bakeries, no water other than tap water is used. Salt brings to flavor, and you can use several types of salts (marine, with herbs, etc.). The bread doesn't have to have salt; many loaves are traditionally bland, and others are brushed with a saline solution when leaving the oven (especially the loaves with little crumbs).

Flour, yeast, water, and salt are essential ingredients you need to make a good bread dough. Before carrying it to the oven comes the work of fermentation, kneading, etc. But as a useful note (and to encourage you to make bread), we must remember that not all the loaves are baked, nor do they need many hours of fermentation and kneading.

The bread can be made in a pan, griddle, casserole, or steamed, and are an excellent option for when you do not want to heat the oven. You can also make many other bread pieces in skillets, such as pita bread, Moroccan bread, bread from North Africa, and various flatbreads. Making bread in a pan is one of the eldest ways of cooking it. If you decide to make bread in the pan, you must choose a good one that keeps the heat well and can cook evenly.

Conclusion

Depending on what kind of home baker you are, bread is either a must-know rite of passage or an intimidating goal you haven't quite worked up the courage to try. This is because bread is a labor-intensive food where slight mistakes can have a big impact on the final result. Most of us rely on store-bought bread, but once you've tasted homemade bread, it's tempting to make your own as often as possible. A bread machine makes the process easier.

Making a loaf of bread feels like a major accomplishment. Why? There are a lot of steps. Mixing, kneading, proofing, resting, shaping, and finally baking.

Do you know how to make bread by hand, so how does the bread machine do it? A bread machine is basically a small electric oven. It fits on a bread pan with a shaft connected to the motor. A metal paddle is attached to the shaft and the dough is kneaded. If you were making bread in a mixer, you would probably use a dough hook, and in some instructions, you will see that the bread machine kneader is called a hook or "blade".

The first thing you do is take out the tin and add the bread dough you made in Step 1. Bread machines can make any kind of bread, whether it's made from normal white flour, whole-wheat, etc. Pop this tin into the axle and program it by selecting the type of bread, which includes options like basic, whole-wheat, multigrain, and so on. There are even cycles specifically for sweet breads; breads with nuts, seeds, and raisins; gluten-free, and bagels. Many models also let you cook jam.

You'll probably see a "dough" mode option, too. You would use that one for pizza. The machine doesn't actually cook anything; it just kneads and then you take out the pizza dough and bake it in your normal oven. If you aren't making pizza dough, the next selections you'll make are the loaf size and crust type. Once those are chosen, press the "timer" button. Based on your other selections, a time will show up and all you have to do is push "start."

After kneading and before the machine begins baking, many people will remove the dough so they can take out the kneading paddles since they often make an indent in the finished bread. The paddles should simply pop out or you can buy a special hook that makes the removal easier. Now you can return the bread to the machine. The lid is closed during the baking process. If it's a glass lid, you can actually see what's going on. You'll hear the paddle spinning on the motor, kneading the dough. It lies still for the rising stage and then starts again for more kneading if necessary. The motor is also off for the proving stage. Next, the heating element switches on, and steam rises from the exhaust vent as the bread bakes. The whole process usually takes a few hours.

There is a lot of work going on in making bread by hand. When you use a machine, this machine does many of the busiest tasks for you. Just add your dough and the baker will start doing his job, giving you time to do other chores or to sit back and relax. As a note, not all bread machines are automatic, so if you want this benefit, you will have to pay a little more money. But for many people, it is worth it.

Bread machines are indeed easy to use. If you can use a crockpot or a microwave, you can use a bread machine. Cycles and other settings like loaf size and color are always clearly marked, and once you do a quick read of your instruction manual, you'll be ready to go. Recipes written for bread makers are also very clear about what settings you need to select, so as long as you follow them, your bread will turn out the way you want.

Ovens require a lot of electricity and when you're making bread, that long bake time can make an impact on your energy bill. They also lose a lot of energy because the oven is much larger than necessary for one loaf of bread. Bread makers are smaller and therefore more efficient.

Freshly baked homemade bread—The list of benefits is seemingly endless; the taste of something homemade is typically superior to anything store bought and there's the added advantage of knowing the ingredients used—nothing artificial in here. This is especially important for those with allergies.

Less mess to clean up as all the ingredients are mixed in one pan, and with the ease of using the controlled settings, there's no kneading or prolonged wait times.

Like with any kitchen gadget, taking care of your bread machine is important for its longevity and safety. These appliances can last for years when they're well-cared for.

What's important is that the outcome will be delicious homemade bread.

CPSIA information can be obtained
at www.ICGtesting.com
Printed in the USA
BVHW061713070621
608939BV00007B/1045

9 781914 136832